Expanding Mathematical Toolbox: Interweaving Topics, Problems, and Solutions

This book offers several topics from different mathematical disciplines and shows how closely they are related. The purpose of the book is to direct the attention of readers who have an interest in and talent for mathematics to engaging and thought-provoking problems that should help them change their ways of thinking, entice further exploration, and possibly lead to independent research and projects in mathematics. In spite of the many challenging problems, most solutions require no more than a basic knowledge covered in a high-school math curriculum.

To shed new light on a deeper appreciation for mathematical relationships, the problems are selected to demonstrate techniques involving a variety of mathematical ideas. Included are some interesting applications of trigonometry, vector algebra and Cartesian coordinate system techniques, in addition to geometrical constructions and inversion in solving mechanical engineering problems and in studying models explaining non-Euclidean geometries.

Expanding Mathematical Toolbox: Interweaving Topics, Problems, and Solutions is primarily directed at secondary school teachers and college professors. The book will be useful in teaching mathematical reasoning because it emphasizes how to teach students to think creatively and strategically and how to make connections between math disciplines. The text also can be used as a resource for preparing for mathematics Olympiads. In addition, it is aimed at all readers who want to study mathematics, gain a deeper understanding and enhance their problem-solving abilities. Readers will find fresh ideas and topics offering unexpected insights, new skills to expand their horizons in math studies, and an appreciation for the beauty of mathematics.

Expanding Mathematical Toolbox: Interweaving Topics, Problems, and Solutions

Boris Pritsker

CRC Press
Taylor & Francis Group
Boca Raton London New York

CRC Press is an imprint of the
Taylor & Francis Group, an **informa** business

A CHAPMAN & HALL BOOK

First edition published 2023
by CRC Press
6000 Broken Sound Parkway NW, Suite 300, Boca Raton, FL 33487-2742

and by CRC Press
4 Park Square, Milton Park, Abingdon, Oxon, OX14 4RN

CRC Press is an imprint of Taylor & Francis Group, LLC

ISBN: 9781032417387 (hbk)
ISBN: 9781032417356 (pbk)
ISBN: 9781003359500 (ebk)

DOI: 10.1201/9781003359500

Typeset in Palatino
by KnowledgeWorks Global Ltd.

To the memory of my beloved parents.

Contents

Contents

Preface

Mathematics isn't a palm tree, with a single long straight trunk covered with scratchy formulas. It's a banyan tree, with many interconnected trunks and branches - a banyan tree that has grown to the size of a forest, inviting us to climb and explore.

William P. Thurston

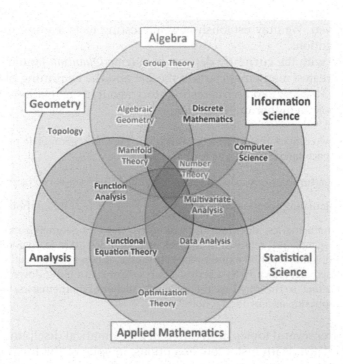

While studying mathematics in middle and then in secondary school, students face arithmetic, algebra, geometry, trigonometry, pre-calculus, and statistics and probability as separate disciplines. A stereotypical view has been to "divide" mathematics into categories that are, by implication, close to disjoint. Don't we apply pure algebraic techniques for solving quadratic or cubic equations? What about solving construction problems in geometry; don't we rely on the geometric properties of figures and use already proven geometric theorems? Moreover, it tends to be that synthetic solutions are usually valued more than algebraic or trigonometric. Certainly, this is true. Many algebraic and geometric problems are efficiently solved without referring to other math disciplines. Some scholars even consider, for instance, trigonometry as the killer of geometrical beauty, meaning that tedious trigonometric modifications sometimes overshadow pure elegant geometrical ways of problem-solving. In fact, it all depends on the problem one is solving, and in most cases, it is up to you to decide what methods and techniques to apply to get the most elegant and short solution. But it is an erroneous view that mathematical branches are not associated with each other and each one exists in its own universe under its own laws. One cannot succeed in mathematical studies without relating math disciplines to each other and broadly incorporating multiple techniques in the problem-solving process. Needless to say, one cannot fully appreciate the inherent beauty of mathematics. Different aspects of the mathematical sciences do not proceed in isolation from one another. NCTM defines mathematical connections in *Principles and Standards for School Mathematics* as the ability to "recognize and use connections among mathematical ideas; understand how mathematical ideas interconnect and build on one another to produce a coherent whole; recognize and apply mathematics in contexts outside of mathematics." Also, an important pedagogical technique is to implement multiple approaches to a problem-solving process and show the interrelations between the fields. Searching for alternative solutions to the same problem and new proofs of the well-known theorems, especially applying different techniques from math disciplines, allows for a better understanding of the principles and connections involved. We may establish new interesting facts leading to generalizations and even new inventions.

In his interview with the currently defunct magazine *Quantum* January/February 1991 issue, one of the greatest mathematicians of the 20th and the beginning of the 21st century, Israel Gelfand (1913 – 2009) told an interesting story about his revelations at age of 15 after reading the textbook on higher mathematics:

> "I took away two remarkable ideas from this book. First, any geometric problem in the plane and in space can be written as formulas...
>
> The second idea turned my worldview upside down. This idea is the fact that there's a formula for calculating the sine: $sin\ x = x - \dfrac{x^3}{3!} + \dfrac{x^5}{5!} - \cdots$. Before this, I thought there are two types of mathematics, algebraic and geometric, and that geometric mathematics is basically "transcendental" relative to algebraic mathematics - that is, in geometry, there are some notions that can't be expressed by formulas. When I discovered that the sine can be expressed algebraically as a series, a barrier came tumbling down, *and mathematics became one*. To this day I see the various branches of mathematics, together with mathematical physics, as a unified whole."

In this book we offer several topics from different mathematical disciplines and show how closely they are related. With just a few exceptions, in spite of the many problems' challenging character, their solutions require no more than a basic knowledge covered in a high school math curriculum. For this reason, as we feel that complex numbers represent a more advanced topic, the discussion of complex numbers is omitted, and we restrict all our studies to the field of real numbers only. The problems are selected to demonstrate techniques that involve a variety of mathematical ideas which can shed new light on a deeper appreciation for mathematical relationships. We go through some interesting applications of geometrical constructions and inversion in solving mechanical engineering problems and in studying models explaining non-Euclidean geometries. We examine algebraic problems efficiently solved with pure geometric techniques and vice versa. We investigate how useful trigonometry applications and vector algebra techniques are in establishing links between algebra, geometry, and number theory. Rene Descartes' invention of the Cartesian coordinate system had the effect of allowing the conversion of geometry into algebra (and vice versa). It also implemented a very powerful tool alternative to conventional techniques for tackling many difficult problems; we cover several such examples in the chapter devoted to this topic. Mathematical inequalities play a fundamental role not just in the development of various branches of mathematics but in mathematical economics, game theory, control theory, operations research, probability and statistics, and mathematical programming. The topic of inequalities is linked to others in mathematics. We go through many challenging problems involving algebraic and geometric inequalities and their applications in problem-solving. The final chapter is fully devoted to the Guess-and-Check technique as a nice alternative to traditional math techniques for solving interesting problems; it provides another insight into establishing sometimes unexpected connections among math disciplines.

The book is primarily directed to secondary school teachers and college professors. It can be used as supplemental reading for algebra, geometry, and undergraduate pre-calculus courses. It should be useful in teaching Mathematical Reasoning courses as well because it emphasizes how to teach students to think creatively and strategically and how to make connections among multiple fields of math. The book also can be used as a resource for preparing for mathematics Olympiads. Evaluating unorthodox approaches to problem-solving and searching the links among math disciplines, often lead to useful results and even new discoveries. The purpose of the book is to direct the attention of readers who have an interest in and talent for mathematics to engaging and thought-provoking problems that should help them change their ways of thinking, entice further exploration, and could lead to independent research and projects in mathematics.

It is not only aimed at those students who are considering a career in mathematics or a related field, but also to all who want to study mathematics with the help of our book, gain a deeper understanding, and enhance their problem-solving abilities. Many of the problems that you will come across in the book seem completely baffling to the untrained person. However, they can be solved using surprisingly simple techniques. In some instances, the results seem to emerge miraculously, out of thin air. We hope the readers will find new ideas and topics that give unexpected insights and perhaps some new skills to expand the horizons in math studies and to get to the essence of the matter: enjoyment of the beauties of mathematics.

About the Author

Boris Pritsker studied mathematics at Kiev State Pedagogical University, Ukraine, and then worked as a math teacher in high schools including a special math-oriented school for gifted and talented students with advanced programs in algebra, geometry, trigonometry, and calculus. In the US, he earned an MBA degree from the Graduate School of Baruch College, City University of New York. He is a licensed CPA in New York State and has been employed by CBIZ Marks Paneth LLC, a New York accounting and consulting firm, where he is a director.

He published numerous articles and problems in mathematical magazines in the former Soviet Union, Unites States, Australia, and Singapore. He is also the author of three internationally acclaimed mathematics books.

1

Beauty in Mathematics

Mathematics, rightly viewed possesses not only truth but supreme beauty.

Bertrand Russel

There is no definition of "beauty" or "elegance". These are very subjective qualities. We may say that it is some combination of qualities that pleases the intellect or moral sense. Perceptions of beauty have varied throughout centuries and in different cultures. Beauty or elegance is often simply a matter of taste. In fact, no concept could ever fully define beauty. What one person perceives as beautiful, somebody else would not consider as such. As the great Persian mathematician, philosopher, and poet Omar Khayyám once said

> In one window looked two. One saw the rain and mud. Other – green foliage ligature, spring, and the sky is blue. In one window looked two.

Would those huge lines of people in front of Leonardo da Vinci's *Mona Lisa* painting in the Louvre Museum or Michelangelo's *David* statue in Galleria dell' Academia define the meaning of beauty? Is it something that impresses the majority of people to the point that it inspires and represents the most distinct and remarkable attractiveness of our souls?

I remember my first visit to Grand Canyon National Park. Our tour guide did not allow anyone from our group to go alone to the lookout point; he insisted that we all came together. It was one simultaneous "Wow!!!" as we got to the observing place. We asked our tour guide why he did not let us go there without him. "I wanted to witness this 'Wow!' reaction", was his response. It was one of those moments that draw upon our strength and consume us with the remarkable and thrilling experience of beautiful scenes of nature. Should this "Wow!" experience serve as a valid definition of beauty?

What about defining beauty and elegance in mathematics?

Don't you have this "Wow!" feeling looking at the examples below?

$$1 \cdot 8 + 1 = 9$$
$$12 \cdot 8 + 2 = 98$$
$$123 \cdot 8 + 3 = 987$$
$$1234 \cdot 8 + 4 = 9876$$
$$12345 \cdot 8 + 5 = 98765$$
$$123456 \cdot 8 + 6 = 987654$$
$$1234567 \cdot 8 + 7 = 9876543$$
$$12345678 \cdot 8 + 8 = 98765432$$
$$123456789 \cdot 8 + 9 = 987654321$$

DOI: 10.1201/9781003359500-1

$$1 \cdot 9 + 2 = 11$$
$$12 \cdot 9 + 3 = 111$$
$$123 \cdot 9 + 4 = 1111$$
$$1234 \cdot 9 + 5 = 11111$$
$$12345 \cdot 9 + 6 = 111111$$
$$123456 \cdot 9 + 7 = 1111111$$
$$1234567 \cdot 9 + 8 = 11111111$$
$$12345678 \cdot 9 + 9 = 111111111$$
$$123456789 \cdot 9 + 10 = 1111111111$$

$$9 \cdot 9 + 7 = 88$$
$$98 \cdot 9 + 6 = 888$$
$$987 \cdot 9 + 5 = 8888$$
$$9876 \cdot 9 + 4 = 88888$$
$$98765 \cdot 9 + 3 = 888888$$
$$987654 \cdot 9 + 2 = 8888888$$
$$9876543 \cdot 9 + 1 = 88888888$$
$$98765432 \cdot 9 + 0 = 888888888$$

Brilliant, isn't it?

And look at this symmetry:

$$1 \cdot 1 = 1$$
$$11 \cdot 11 = 121$$
$$111 \cdot 111 = 12321$$
$$1111 \cdot 1111 = 1234321$$
$$11111 \cdot 11111 = 123454321$$
$$111111 \cdot 111111 = 12345654321$$
$$1111111 \cdot 1111111 = 1234567654321$$
$$11111111 \cdot 11111111 = 123456787654321$$
$$111111111 \cdot 111111111 = 12345678987654321$$

How about the amazing formulas discovered by one of India's greatest mathematicians Srinivasa Ramanujan (1887–1920):

$$\sqrt{1 + 2\sqrt{1 + 3\sqrt{1 + 4\sqrt{1 + \cdots}}}} = 3$$

$$\sqrt{6 + 2\sqrt{7 + 3\sqrt{8 + 4\sqrt{9 + \cdots}}}} = 4$$

$$1 + \frac{1}{1 \cdot 3} + \frac{1}{1 \cdot 3 \cdot 5} + \frac{1}{1 \cdot 3 \cdot 5 \cdot 7} + \cdots + \cfrac{1}{1 + \cfrac{1}{1 + \cfrac{2}{1 + \cfrac{3}{1 + \cfrac{4}{1 + \cdots}}}}} = \sqrt{\frac{\pi e}{2}}.$$

What do we mean when we call a solution to a problem an "elegant and beautiful solution"? Surely, there is no such definition, it is very subjective. Prolific British mathematician Arthur Cayley (1821–1895) proclaimed that "As for everything else, so for mathematical theory: beauty can be perceived but not explained". What I have discovered is that, first of all, beauty is simple. Moreover, it should be something vivid and easy to comprehend. In many cases, an elegant solution assumes utilizing of some efficient trick or strategy significantly simplifying a solution. If one stands in awe and feels impressed and amused at the same time, then most likely, a problem was solved in unusual and engaging manner, and we can agree that it was an elegant solution.

How about this cute brain teaser? Dare we call it a problem, but isn't it fun?

How do you write 4 in between of 5?

The answer:

$$F(IV)E.$$

In our introductory chapter, we are going to look into several interest-provoking challenges, solutions of which are unusual and not typically anticipated. They have nothing in common among them other than their charming and delighting solutions.

We will start with a few cute logical puzzles resembling the one above, leaving more advanced challenges for the second part of this chapter.

Tackling many interesting problems, it is critical to fully understand what the problem is about and what is asked. Sometimes, an unexpected missing outcome surprises you when you think that the problem is solved, while there are some missing solutions. We strongly encourage the readers to try all the problems offered below *before* reading their solutions. Have fun!

PROBLEM 1.1.

Offered by S. Kostin in *Квант* (in Russian) magazine, 2017, #12.

Number 73 is formed with eight matches of the same length, as in the figure below. Can you rearrange two matches to get a square?

SOLUTION.

The first solution is very easy to find – move matches 1 and 6 as it is shown below:

What about the second solution? Can you find it? Would you be surprised to see it as 16 below:

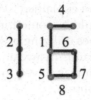

The goal was to get the square. Since $16 = 4^2$, then 16, as depicted in the figure above, is our second solution. The ambiguity was in the word "square". But 16 is a square of 4, so this is unexpected, but a valid solution to the problem as well.

PROBLEM 1.2.

Is it possible to arrange four equilateral triangles using six matches of the same length (assuming that every one of these matches represents a side of a triangle)?

<u>SOLUTION</u>.

What I found interesting about this problem is that one naturally starts trying to form the equilateral triangles from 6 matches placing them on a table, i.e., working the problem out in a two-dimensional plane. After several futile attempts, one arrives at the inevitable conclusion–this is impossible; the problem has no solutions! Indeed, it is not solvable in a plane. But no such restriction is stated in the problem. As soon as we expand into the three-dimensional space, we should be able to construct a *regular tetrahedron*. A regular tetrahedron consists of four equilateral triangles, so it satisfies the conditions of our problem!

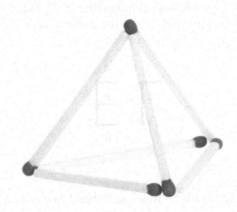

PROBLEM 1.3.

We can easily form two rectangles from 18 matches (all matches are of the same length), such that one has the area that is twice as big as the area of the other one, as it is depicted in the figure below.

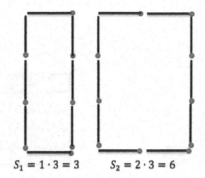

$$S_1 = 1 \cdot 3 = 3 \qquad S_2 = 2 \cdot 3 = 6$$

Can you rearrange these 18 matches to form two quadrilaterals such that the area of one of them is three times greater than the area of the other one?

SOLUTION.

The goal will be achieved by making a rectangle from 6 matches and a parallelogram from 12 matches as shown in the figure above. The altitude in our parallelogram should be 1.5 of the length of the given matches. Then the area of a rectangle is 2 square units, and the area of a parallelogram is $1.5 \cdot 4 = 6$ square units, i.e., it is three times greater, as it was demanded in the conditions of the problem.

One of the greatest mathematicians ever, Carl Friedrich Gauss (1777–1855), is also known as one of the most elegant mathematicians of all time. Let's recall a well-known mathematical legend from his childhood that you've probably heard before.

Carl Gauss has possessed enormous math talents from early childhood. A 7- to 8-year-old boy (there are variations of the date of this incident in math literature) was always the first to quickly solve every problem given by a math teacher to his class. Trying to keep this student quiet and busy, a math teacher offered him to add all integers from 1 to 100 and ordered him to not bother him until the result is derived. The teacher hoped that this tedious assignment should take considerable time to be completed. However, in a few minutes, the answer was delivered to him. Instead of going through adding all integers one by one, the young genies suggested adding the sums of pairs of numbers equidistant from both ends of the sum:

$$1 + 2 + \cdots + 99 + 100 = (1 + 100) + (2 + 99) + \cdots + (50 + 51) = 101 \cdot 50 = 5050.$$

Amazing! Isn't it?

To extend the trick of adding the sums of numbers equidistant from both ends of the sum of any n integers, we conclude that the sum of natural numbers from 1 through n is expressed by the formula: $1 + 2 + \cdots + n = \dfrac{n(n+1)}{2}$.

It has a vivid geometrical illustration.
Let's consider, for instance, $n = 5$.

$$1+2+3+4+5$$

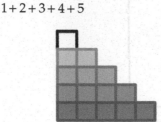

$$1+2+3+4+5 = \frac{5 \cdot 6}{2} = 15$$

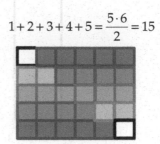

Interesting is that by applying Gauss' adding technique one can get a general formula for an arithmetic series of n terms: $S = \frac{(a_1 + a_n)}{2} \cdot n$ (here a_1 is the first term and a_n is the last term).

PROBLEM 1.4.

Prove that for any $x > 0$, $x + \frac{1}{x} \geq 2$.

SOLUTION.

This famous inequality is easy to prove algebraically considering the difference $x + \frac{1}{x} - 2$ and completing a square:

$x + \frac{1}{x} - 2 = \frac{x^2 + 1 - 2x}{x} = \frac{(x-1)^2}{x} \geq 0$. The last fraction is non-negative because we know that $x > 0$ and $(x-1)^2 \geq 0$ as a square of a number (this is always a non-negative number).

There is a vivid geometrical illustration of this property as well.

Consider four rectangles with sides x and $\frac{1}{x}$ arranged as in the figure below to form a square. The length of the side of the square is $\left(x + \frac{1}{x} \right)$. So, its area equals $\left(x + \frac{1}{x} \right)^2$.

On the other hand, clearly, the area of this square is greater or equal than the sum of the areas of our four rectangles, which is $4 \cdot x \cdot \frac{1}{x} = 4 = 2^2$, and can be interpreted as the area of a square with the side 2.

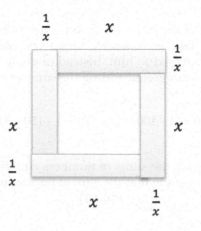

Hence, we conclude that since $\left(x+\dfrac{1}{x}\right)^2 \geq 4$, then, obviously, $x+\dfrac{1}{x} \geq 2$, as it was required to be proved.

We may use this geometrical illustration to make an even stronger argument. The area of our square is greater than the area of a square with the side 2 by the area of a small square inside the big one (its side's length is $x-\dfrac{1}{x}$). Indeed,

$$\left(x+\frac{1}{x}\right)^2 - \left(x-\frac{1}{x}\right)^2 = x^2 + 2 + \left(\frac{1}{x}\right)^2 - x^2 + 2 - \left(\frac{1}{x}\right)^2 = 4.$$

Let's now go over a few geometrical challenges that have elegant and vivid solutions.

PROBLEM 1.5.

An artificial lake has the shape of a square. Big oak trees were planted at each vertex of the lake. Is it possible to expand the lake preserving its shape of a square but doubling its area without removing oak trees, so they stand by the banks of the expanded lake (edges of the new square)?

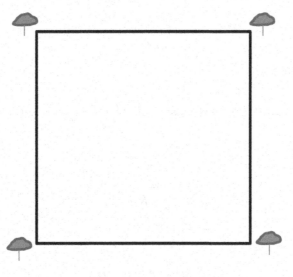

<u>**SOLUTION.**</u>

Assume the oak trees are positioned at points A, B, C, and D of the square $ABCD$. Our goal is to construct a square such that its area is double of the area of $ABCD$, and points A, B, C, and D are on its sides. The easiest solution will be to draw four triangles on the sides of our square congruent to respective triangles in which $ABCD$ is cut by its diagonals (O is the point of intersection of the diagonals AC and BD): $\triangle AMB \cong \triangle AOB$, $\triangle BNC \cong \triangle BOC$, $\triangle CKD \cong \triangle COD$, and $\triangle AED \cong \triangle AOD$. The area of the newly formed square $MNKE$ is twice of the area of $ABCD$ because it consists of $ABCD$ itself and four congruent triangles such that area of each one is $\dfrac{1}{4}$ of the area of $ABCD$. We did not move any of the trees while keeping each of them on the sides of a newly expanded lake and managed to double the area of the lake while preserving its shape. Therefore, all of the conditions are satisfied, and our problem is solved.

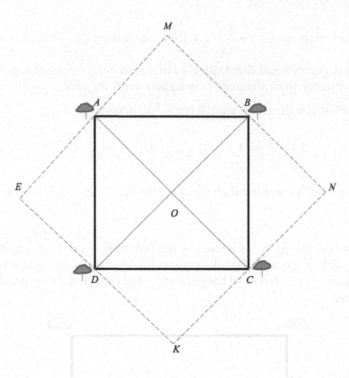

PROBLEM 1.6.

In a convex hexagon $ABCDEM$, $AM = ED = CB$, $\angle E = \angle C = \angle A$, and $\angle D = \angle B = \angle M$. Prove that $AB = ME = DC$.

SOLUTION.

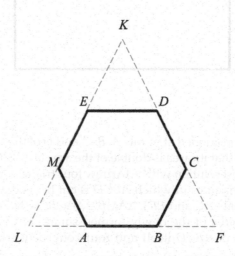

By extending the sides of our hexagon till their intersection at L, K, and F we form a triangle LKF. It is very easy to prove (we leave this to the readers) that according to the given conditions, triangles ALM, EKD, and CFB are congruent by Angle-Side-Angle property.

It implies that the three respective angles by the vertices of triangle *LKF* are congruent, $\angle L = \angle K = \angle F$, and therefore, this is an equilateral triangle. Hence, $LK = KF = LF$. Since we are given that $AM = ED = CB$ and the pairs of the respective sides in congruent triangles *ALM*, *EKD*, and *CFB* are congruent, it follows that $AB = ME = DC$, as wanted.

Introduction of an auxiliary triangle *LKF* enabled us to get very short, vivid, and elegant proof of the problem's assertion!

While solving many difficult problems, we look for some auxiliary variables or auxiliary elements as we just did in the previous problem; or even supplemental easier problems, which help us not just make a solution manageable but elegant as well.

In many problems, it is very important to identify the most critical conditions or attributes and take advantage of properties associated with them to simplify your solution.

PROBLEM 1.7.

Solve the equation:

$$\left(x^2 + 2x - 3\right)^3 + \left(3x^2 - 4x + 1\right)^3 = \left(4x^2 - 2x - 2\right)^3.$$

SOLUTION.

This equation looks tough. Conventional strategy cubing each expression with further simplifications of the resulting expressions most likely would lead nowhere. A clever approach to any problem assumes careful review of what we are dealing with. If we notice and identify some critical data in a problem (some attributes that distinguish a problem from other similar problems), then it may help us to find a relatively simple and elegant solution that is enabled by this discernable condition.

Here it is not hard to realize that the bases of the exponents on the left-hand side sum up to equal the base of the exponent on the right-hand side:

$$x^2 + 2x - 3 + 3x^2 - 4x + 1 = 4x^2 - 2x - 2.$$

We have the sum of cubes on the left-hand side of the equation. Should we recall a general formula $(u + v)^3 = u^3 + v^3 + 3uv(u + v)$, and utilize it in our solution?

Even though it looks like a weird idea to go from a single-variable equation to a multi-variable equation, we can try introducing two new variables and see if this would simplify our solution.

Let $u = x^2 + 2x - 3$ and $v = 3x^2 - 4x + 1$. We can now rewrite the equation as $u^3 + v^3 = (u + v)^3$, from which it follows that $3uv(u + v) = 0$, or equivalently, $uv(u + v) = 0$. Substituting back these expressions for x gives

$$\left(x^2 + 2x - 3\right)\left(3x^2 - 4x + 1\right)\left(4x^2 - 2x - 2\right) = 0.$$

Our original equation, therefore, is simplified to the last equation, the solution of which comes to solving three quadratic equations:

$$x^2 + 2x - 3 = 0, \text{ or } 3x^2 - 4x + 1 = 0, \text{ or } 4x^2 - 2x - 2 = 0.$$

Using Viète's formulas, the solutions of the first equation are $x_1 = 1$, $x_2 = -3$.

Solutions of the second equation are $x_3 = 1$, $x_4 = \dfrac{1}{3}$.

Finally, solving the third equation $2x^2 - x - 1 = 0$, gives $x_5 = 1$, $x_6 = -\dfrac{1}{2}$.

Answer: $1, \dfrac{1}{3}, -3, -\dfrac{1}{2}$.

Don't you agree that non-conventional strategy of introducing two auxiliary variables allowed us to get an elegant and easy solution?

PROBLEM 1.8.

Solve the equation $\cos x + 2\cos 2x + \cdots + n\cos nx = \dfrac{n^2 + n}{2}$.

SOLUTION.

Observing that for any x, $|\cos x| \le 1$, we obtain that

$$|2\cos 2x| \le 2, |3\cos 3x| \le 3, \ldots, |n\cos nx| \le n.$$

It follows that our left-hand side does not exceed the right-hand side. Indeed,

$$\cos x + 2\cos 2x + \cdots + n\cos nx \le 1 + 2 + \cdots + n = \frac{n(n+1)}{2} = \frac{n^2+n}{2}.$$

Therefore, the equality is possible only when each addend on the left-hand side attains its maximum value, i.e., $\cos x = 1$, $\cos 2x = 1$, ..., $\cos nx = 1$. Solving each of these equations we get that respectively, $x = 2\pi n$, $n \epsilon Z$; $x = \pi k$, $k \epsilon Z$, ..., $x = \dfrac{2\pi m}{n}$, $m \epsilon Z$. The common solutions of all these equations are expressed as $x = 2\pi n$, $n \epsilon Z$.

Answer: $x = 2\pi n$, $n \epsilon Z$.

"An elegant proof just hits you between your eyes and fills your heart with joy", said American mathematician Irving Kaplansky (1917–2006). To come up with such a proof, not only does one need to possess a strong mathematical background and deep knowledge of the subject matter but be creative and unbounded in applying such knowledge in tackling interesting math challenges. In many instances when solving difficult problems, we restrict ourselves to only theorems, laws, or formulas (tools in our possession) seemingly directly related to our problem. Conventional strategies assume that in solving geometrical problems we tend to think in a "geometric direction", and in solving algebraic problems, we first try to incorporate algebraic techniques. The ability to expand the horizons and see the broader picture by applying the links among various math's (not even necessarily purely math!) disciplines often provides great results in finding elegant and beautiful solutions.

Let's demonstrate, for instance, how unexpectedly useful the laws of mechanics might be in solving certain geometrical problems.

Archimedes' Law of the Lever states that

> *Magnitudes are in equilibrium at distances reciprocally proportional to their weights.*

The Law of the Lever

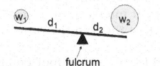

fulcrum

$$w_1 \times d_1 = w_2 \times d_2$$

We assume that the readers are familiar with this law, as well as with the concept of the *center of mass* of a distribution of mass in space, the so-called *balance point*, the unique point where the weighted relative position of the distributed mass sums to zero. If you don't know about this concept, many books in advanced Euclidian geometry will fill you in.

The center of mass is the exact center of all the material an object is made of. An object's center of mass is the point at which it can be balanced. It is widely applicable in solving various problems in mechanics. But it can be very efficient in approaching geometrical problems as well.

In my mind one of the most elegant proofs that the three medians of a triangle are concurrent at the point called the *centroid* or *center of mass* of a triangle and that it is $\frac{2}{3}$ of the way from a vertex to the opposite midpoint is related to applying *Archimedes' Law of the Lever* and the concept of the center of mass of the system of material points.

PROBLEM 1.9.

Prove that in any triangle the three medians (median is a segment connecting the vertex with the midpoint of the opposite side) intersect at one point and this point divides each median in a 2:1 ratio.

SOLUTION.

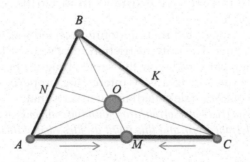

Consider triangle *ABC* and draw its medians *AK*, *BM*, and *CN*. Let's put the same weight of 1 pound at each vertex of our triangle *ABC*. According to *Archimedes' Law of the Lever*, the center of mass of two material points *A* and *C* is located at midpoint *M* of the segment *AC*. Shifting the weights of 1 pound from *A* and *C* to *M*, we will concentrate 2 pounds at *M* not affecting the overall balance of our system of material points. This allows us now to consider the equivalent system consisting of two material points, *B* with the weight of 1 pound and *M* with the weight of 2 pounds. Applying *Archimedes' Law of the Lever* to these two material points, we conclude that its center of mass *O* lies on *BM*, and the distance from *O* to *B* is twice the distance from *O* to *M*. Selecting other medians *AK* and *CN*, we would make the exact same observations. Since a system of material points has a unique center of mass, it implies that indeed *O* is the single point of intersection of three medians (*O* lies on each of the medians) and it divides each of them in the ratio 2:1 counting from a vertex of a triangle, the sought-after result!

The same logic can be extended for the case of four points such that any three of them do not lie on the same straight line. By connecting each of these randomly selected four points with the centroid of a triangle that has its vertices at the other three points we can easily obtain that these four segments intersect at one point and this point divides each of them in the ratio of 3:1.

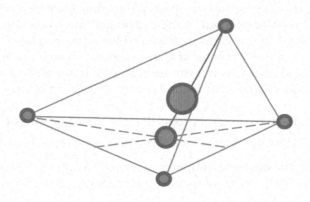

Indeed, by shifting the weights of 1 pound from three vertices of a triangle to its center of mass (centroid), we will get the system of two material points that consists of the mentioned centroid with the weight of 3 pounds and the fourth point with the weight of 1 pound. Applying *Archimedes' Law of the Lever* to these two material points, we obtain that its center of mass is lying on a straight line connecting them, and the distance from this point to the fourth point is three times the distance from it to the centroid of a triangle with vertices at other three points.

Nothing in our analysis suggested that randomly selected four points have to lie in the same two-dimensional plane. The only restriction imposed on them was that any three points should not be collinear (not lie on the same straight line). Launching into three-dimensional space, this brings us to a very interesting property of a tetrahedron that is analogous to a property of a triangle's medians. It asserts that:

Segments connecting the vertices of a tetrahedron with the centroids of the opposite faces intersect at one point (centroid of a tetrahedron), and this point divides each of them in the ratio of 3:1 counting from a vertex.

The ideas similar to those used above enable us to find an unexpected and elegantly brief solution to the following Problem 1.10.

PROBLEM 1.10.

Points K and L lie on the sides AB and BC of the triangle ABC and divide its sides respectively in ratios $\frac{AK}{KB} = \frac{1}{2}$ and $\frac{BL}{LC} = \frac{2}{5}$. Find $\frac{AS}{SL}$ and $\frac{CS}{SK}$.

SOLUTION.

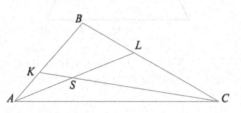

Let's put the following weights at each vertex of the triangle ABC:

$m_A = 10$ pounds at vertex A,

$m_B = 5$ pounds at vertex B,

$m_C = 2$ pounds at vertex C.

According to the conditions of the problem, the center of mass of two material points A and B is located at point K on the segment AB such that $\frac{AK}{KB} = \frac{1}{2}$ and respectively, the center of mass of three material points A, B, and C lies on the segment CK. Also, the center of mass of points B and C is located at point L on the segment BC such that $\frac{BL}{LC} = \frac{2}{5}$ and respectively, the center of mass of three material points A, B, and C (this is the same point as found in the first observation!) lies on the segment AL. It implies that the center of mass of three material points A, B, and C is the point of intersection of CK and AL, point S. Finally, applying *Archimedes' Law of the Lever* we obtain that

$\frac{CS}{SK} = \frac{m_A + m_B}{m_C} = \frac{10+5}{2} = \frac{15}{2}$ and $\frac{AS}{SL} = \frac{m_B + m_C}{m_A} = \frac{5+2}{10} = \frac{7}{10}$, which are the answers to

the problem.

There are many instances when several relatively simple and elegant proofs of geometric problems are based on engaging algebraic and trigonometric techniques. Such solutions look beautiful and seem deceptively easy to find.

PROBLEM 1.11.

Prove that the area of an isosceles triangle is not greater than $\frac{2}{3}$ of the square of the median dropped to one of its equal sides.

SOLUTION.

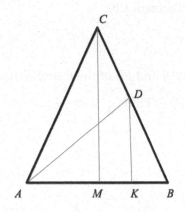

In isosceles triangle ABC ($AC = BC$), AD is the median dropped to side BC. Drawing $CM \perp AB$ and $DK \perp AB$ we obtain similar triangles CMB and DKB in which

$$DK = \frac{1}{2}CM \text{ and } MK = \frac{1}{2}MB = \frac{1}{4}AB.$$

It follows that $AK = AM + MK = \frac{1}{2}AB + \frac{1}{4}AB = \frac{3}{4}AB.$

Letting $AD = c$, $AK = x$, $DK = y$ we can express the area of $\triangle ABC$ (it is equal to half of the base times height) as

$$S_{ABC} = \frac{1}{2}AB \cdot CM = \frac{1}{2} \cdot \left(\frac{4}{3}AK\right) \cdot (2DK) = \frac{4}{3}xy. \tag{1.1}$$

Applying the Pythagorean Theorem to right triangle AKD gives $AK^2 + DK^2 = AD^2$, or in our nominations,

$$x^2 + y^2 = c^2. \tag{1.2}$$

Using the well-known algebraic formula $(x+y)^2 = x^2 + y^2 + 2xy$ and the fact that a square of a number is always non-negative, we observe that $x^2 + y^2 \geq 2xy$. Next, substituting c^2 for $x^2 + y^2$ from (1.2), we see that $c^2 \geq 2xy$, or equivalently, $xy \leq \frac{c^2}{2}$. Finally, turning back to (1.1), we now easily arrive at $S_{ABC} = \frac{4}{3}xy \leq \frac{4}{3} \cdot \frac{c^2}{2} = \frac{2}{3}c^2 = \frac{2}{3}AD^2$, which is the desired result.

Another nice and easy proof of this relationship can be found through the application of trigonometric techniques. This next solution seems even more attractive to me.

Let's use the same figure as above and denote $\angle BAD = \alpha$. Then for the right triangle AKD, $x = c \cdot \cos\alpha$ and $y = c \cdot \sin\alpha$. Therefore, using (1.1) from the above and the trigonometric formula $\sin 2\alpha = 2\sin\alpha \cdot \cos\alpha$ gives

$$S_{ABC} = \frac{4}{3}xy = \frac{4}{3}c \cdot \cos\alpha \cdot c \cdot \sin\alpha = \frac{4}{3} \cdot \frac{c^2}{2}\sin 2\alpha = \frac{2}{3}c^2 \sin 2\alpha.$$

Recalling that the range of the function sine is all real numbers not exceeding 1 in absolute value, we conclude that $S_{ABC} \le \frac{2}{3}c^2$, which is the relationship we set to develop.

Taking another look at our problem, it is natural to pose a question under what conditions the area of an isosceles triangle *is equal* to $\frac{2}{3}$ of the square of the median dropped to one of its equal sides. Isn't it an interesting problem in itself?

Obviously, the equality $S_{ABC} = \frac{2}{3}c^2$ is attained when $\sin 2\alpha = 1$. So, it suffices to solve the trigonometric equation $\sin 2\alpha = 1$ to see that this is possible for $\alpha = 45°$. In other words, $S_{ABC} = \frac{2}{3}c^2$ holds only for an isosceles right triangle. Taking advantage of the trigonometric idea of the solution to the original problem, we devised a surprisingly simple and straightforward solution of the "by-product" problem as well!

In some geometrical problems auxiliary constructions play a crucial role not just in simplifying the solution but in making it vivid and elegant, as is demonstrated in the final problems of our introductory chapter.

PROBLEM 1.12.

In triangle ABC, AK and CP are the altitudes, O is the midpoint of AC, and this triangle has a 60° angle at B, $\angle B = 60°$. Prove that POK is equilateral triangle.

<u>SOLUTION</u>.

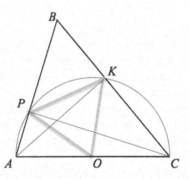

It is given that in triangle ABC, $AK \perp BC$ and $CP \perp AB$, therefore four points A, P, K, and C lie on the circle with center at O and radius $r = OA = OC$. Indeed, if we draw such a circle, then the right angles APC and AKC are subtended by diameter AC, and by the Inscribed angle theorem (see the appendix) it follows that both are inscribed angles, i.e., P and K lie on this circle. Therefore, $OP = OK$ as radii, and we conclude that triangle POK is isosceles. Next, considering right triangle AKB ($\angle K = 90°$) and recalling that $\angle B = 60°$, gives us $\angle BAK = 90° - 60° = 30°$. Clearly, $\angle PAK = \angle BAK = 30°$, and noticing that $\angle PAK$ is the inscribed angle subtended by the same chord PK as the respective central angle POK, by the Inscribed angle theorem, we get that $\angle POK = 2\angle PAK = 2 \cdot 30° = 60°$. This makes the isosceles triangle POK equilateral, as it was required to be proved. By drawing an auxiliary circle, we shifted the focus from our triangle to a different figure, a circle, which enabled us to use its properties to get an easy and elegant solution.

PROBLEM 1.13.

Prove that the area of the regular octagon equals to the product of its two diagonals that have the greatest and the smallest length.

SOLUTION.

The easiest way to solve this problem is to introduce an auxiliary rectangle $ABCD$, as it is shown in the figure below. We can see that triangle $A_8A_1A_2$ consists of two right triangles congruent to triangles A_8AA_1 and A_2BA_1. The same is true for triangle $A_6A_5A_4$ which consists of two right triangles congruent to triangles A_6DA_5 and A_4CA_5.

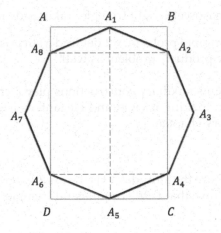

Observing that four triangles $A_8A_1A_2$, $A_2A_3A_4$, $A_4A_5A_6$, and $A_6A_7A_8$ are congruent to each other by Side-Angle-Side property, we get that indeed, the area of the rectangle $ABCD$ is equal to the area of the regular octagon $A_1A_2A_3A_4A_5A_6A_7A_8$; they both consist of the same common part, octagon restricted by the rectangle's sides and segments A_8A_1, A_1A_2, A_4A_5, and A_5A_6, and triangles of the same areas, as explained above. The area of $ABCD$ is calculated as the product of its sides, AB by BC, where each equals respectively to the smallest and the greatest diagonal of the regular octagon. Therefore, we arrive at the desired result. The beauty and benefit of the auxiliary rectangle are manifested in the simplicity of the analysis used to prove the problem's assertion.

Once again, there is no strict definition of beauty. Someone may see this introductory chapter as a hodgepodge of problems, but we hope that these problems are enchanting and should not only generate interest and excitement among readers but demonstrate how useful and enlightening the alternative viewpoints on math problems can be. The conventional strategies tackling interesting challenges are not the best ones in certain cases. Moreover, creatively applying different approaches to solving math problems allows for the devising of elegant and beautiful solutions. The links between seemingly unrelated topics in math disciplines always have great appeal. Making use of constructions with classic Euclidean tools as building bricks in solving various construction puzzles and finding their applications in real-life mechanical problems, solving algebraic problems applying geometrical interpretations and vice versa, solving geometrical problems with help of algebraic analysis and algebraic manipulations, employing trigonometrical identities and properties of trigonometrical functions as connecting links between various math disciplines, utilizing calculus techniques in simplifying difficult problems solutions, applying graphs and geometrical illustrations through vector algebra techniques and the Cartesian coordinate plane techniques – all these and other topics will be examined throughout the subsequent chapters. The solutions we'll discuss are not the only ones possible, nor are they always the shortest. But they are certainly beautiful (in our mind) and we hope you will enjoy them. While you will be working on subsequent chapters, try to solve each problem on your own before looking at the solution. Even if you don't succeed, thinking about the question is certainly beneficial, and you'll perhaps have a greater appreciation of the discussed techniques.

Follow along!

2

Euclidean Constructions

It is the glory of geometry that from so few principles, fetched from without, it is able to accomplish so much.

Sir Isaac Newton

Geometrical construction problems are invaluable in mathematical thinking development, reinforcing originality and inventiveness. Not to mention geometrical education issues alone, any construction problem presents a great model of the whole problem-solving process. These problems are ideal in teaching how to analyze the given conditions and question asked (required construction to perform), make a plan for a solution, carry out such a plan (do the constructions steps), validate each step in the construction process, look for rigorous proof of the achieved results, explore how many solutions a problem may have, discuss alternative solutions, and search for generalizations.

Euclidean constructions are constructions of lengths, angles, and geometric figures using only a straightedge (unmarked ruler) and a compass.

The way the ancient Greeks represented numbers was cumbersome, in comparison with modern nominations. They used similar principles to the decimal numerical system of our days, but to form numbers, instead of the digits from 1 to 9, they used the first nine letters of the old Ionic alphabet from alpha to theta. Instead of reusing these numbers to form multiples of the higher powers of ten, each multiple of ten from 10 to 90 was assigned its own separate letter from the next nine letters of the Ionic alphabet, and so forth. Perhaps as a consequence, they did arithmetic geometrically. The prominent Greek mathematician regarded as the "father of geometry", Euclid of Alexandria (mid-4th century BC–mid-3rd century BC), formulated in the first book of his "Elements" five postulates (axioms) for plane geometry, stated in terms of constructions (as translated by Thomas Heath):

1. To draw a straight line from any point to any point.
2. To produce (extend) a finite straight line continuously in a straight line.
3. To describe a circle with any center and distance (radius).
4. That all right angles are equal to one another.
5. *The parallel postulate*: That, if a straight line falling on two straight lines make the interior angles on the same side less than two right angles, the two straight lines, if produced indefinitely, meet on that side on which the angles are less than two right angles.

Clearly, his axioms are closely related to the tools he used for constructions. Up to these days, when we talk about classic Euclidean construction tools, we mean a straightedge

DOI: 10.1201/9781003359500-2

and compass. These tools suffice to perform many constructions. In particular, you can draw a line parallel to a given line through a given point, construct a line perpendicular to a given line through a given point, and so on. The natural question is – why are we restricted in performing our constructions only to these two instruments? It is important to understand that Euclid intended to build his geometry as an axiomatic system, in which all theorems (statements that need to be proved) are derived from his simple axioms. So, as axioms let us prove everything with a minimum assumptions, compass and straightedge should allow us to construct figures with a minimum of tools. Postulates 1 and 2 relate to constructing straight lines given two distinct points. Postulate 3 relates to constructing a circle given a center and radius. These are the simplest of Euclidean constructions, and in fact, the only basic constructions that one needs to solve any geometric construction problem. Indeed, any straight line is defined by two distinct points, and we use a straight-edge to draw such a line; a circle is determined by its center and the length of its radius, and it can be drawn with the use of a compass. That's all we need. The construction of new points comes from the intersection of two lines, two circles, or a line and a circle.

We assume that readers are well familiar with simple standard constructions with a compass and straightedge that are covered in secondary school geometry curriculum. So, in most cases, we will omit the detailed explanations of every construction step. We will rather concentrate on the important geometrical properties applied in our solutions and on problem-solving strategies on how to get to the desired result.

Solving a construction problem, it is important to find the most efficient solution and do the least number of construction steps. In some problems, it is even clearly stated in a problem's conditions, as in the following problem B284 offered in currently defunct magazine *Quantum*, January/February 2000 issue.

PROBLEM 2.1.

A line l and a point A are given on a plane. Using a compass and a straightedge, construct a perpendicular to l through point A, drawing not more than three lines and circles (the desired perpendicular counts as the last line drawn). Consider two cases, when point A does not lie on line l and when it does.

SOLUTION.

Case 2.1. Point A does not lie on line l.

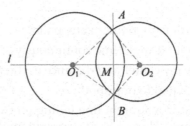

We pick two random points O_1 and O_2 on line l and draw two circles, one with center O_1 and radius O_1A and the second circle with center O_2 and radius O_2A. Clearly, A is the first point where they meet; let's denote B as the second point of their intersection. Then AB is the desired line perpendicular to l.

Let's prove that indeed, $O_1O_2 \perp AB$ (which is the same as to prove that $l \perp AB$).

We observe that triangles O_1AO_2 and O_1BO_2 are congruent by three sides (O_1O_2 – common side, $O_1A = O_1B$ as radii of the first circle, and $O_2A = O_2B$ as radii of the second circle). Therefore, all respective angles in these triangles are congruent as well, and particularly, $\angle AO_1O_2 = \angle BO_1O_2$. This implies that O_1O_2 is the angle bisector in isosceles triangle AO_1B. Therefore, O_1O_2 is the median and the altitude as well, which means that $O_1O_2 \perp AB$, as it was required to be proved. Since we constructed only two circles and one straight line, the conditions of the problem are satisfied, and we are done with Case 2.1.

Case 2.2. Point A lies on line l.

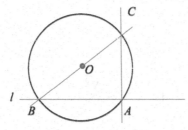

Here we will draw an arbitrary circle whose center O lies off the given line l and which passes through A. Denote B the second point of intersection of the circle and l and draw BO till its intersection with our circle at point C. Final step is to draw CA. This is the desired perpendicular to l. This conclusion immediately follows from the fact that inscribed angle BAC is subtended by diameter BC. Hence, this is a right angle, which proves that $l \perp AC$. In this case we constructed one circle and two straight lines. So, again, the conditions of the problem are satisfied, and our solution is complete.

As we have two segments, one of length x and the other of length y, and a unit length of 1, and define arithmetic operations of addition, subtraction, division, multiplication, and square rooting operation, we can introduce the notion of "constructible numbers". Numbers constructed using straightedge and compass are called constructible numbers.

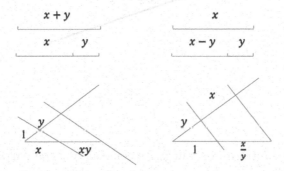

Students rarely relate studied sets of numbers (natural, integers, rational, irrational, real, and complex) to their geometrical expressions. However, in fact, many construction problems are reduced to performing constructions of some lengths expressed by either constructible or non-constructible numbers. So, it is a very important issue. The notion of length became one of the first bridges between algebra and geometry. For ancient Greeks, problems related to "constructible numbers" did not represent some theoretical issues, but rather practical life important issues.

Whereas the question of how to construct the segment with length expressed as a natural or rational number usually presents an easy task to do, then the same question about irrational numbers creates a real challenge. The unsuccessful attempts to measure $\sqrt{2}$ by means of rational line segments puzzled and troubled ancient Greek mathematicians. These failures significantly contributed to mathematical research and development of the study of irrational numbers.

For some irrational numbers construction can be done by applying the Pythagorean Theorem. For instance, to draw a segment where the length equals $\sqrt{2}$, one needs to construct an isosceles right triangle with each leg of the length of 1. The hypotenuse of this triangle has length that equal $\sqrt{1^2 + 1^2} = \sqrt{2}$.

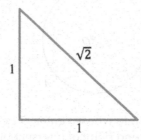

In a similar fashion one can draw many other irrational numbers looking for a hypotenuse of a right triangle with legs of different lengths. For instance, let's construct $\sqrt{22}$.

First, we construct a right triangle with legs 2 and 3. Its hypotenuse is $\sqrt{3^2 + 2^2} = \sqrt{13}$. Second, we construct another right triangle using the already constructed segment with length $\sqrt{13}$ as one of its legs and the second leg with the length of 3. The hypotenuse of that triangle has length $\sqrt{(\sqrt{13})^2 + 3^2} = \sqrt{13 + 9} = \sqrt{22}$.

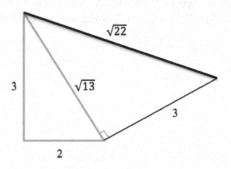

Let's consider one more intriguing strategy for constructing some irrational numbers.

PROBLEM 2.2.

Having a segment with length x, construct a segment with length $x\sqrt{2}$.

<u>**SOLUTION.**</u>

Rewriting $x\sqrt{2}$ as $\sqrt{2x \cdot x}$, we can see that our problem is reduced to constructing the geometrical mean of two segments with lengths x and $2x$.

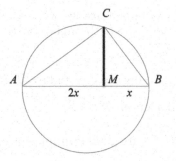

Placing two segments with lengths of $2x$ and x next to each other, we form a segment $AB = 3x$ (where $AM = 2x$, $MB = x$). Draw a circle with center at mid-point of AB and with radius $r = \frac{1}{2} AB$. The next and final step will be to draw at M perpendicular to AB till its intersection with circle at C, $MC \perp AB$. $\angle ACB = 90°$ as the angle inscribed in a semicircle. Hence, from similarity of right triangles AMC and CMB (all respective angles are congruent) it follows that $\frac{AM}{MC} = \frac{MC}{MB}$, which leads to $MC = \sqrt{AM \cdot MB} = \sqrt{2x \cdot x} = x\sqrt{2}$.

In a general case, considering a unitary segment (its length is 1) and a segment with length x, we can now easily construct a segment with the length of $\sqrt{x}$.

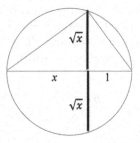

The fact that in a right triangle the altitude on the hypotenuse is equal to the geometric mean of line segments formed by altitude on the hypotenuse, and the converse assertion, that is also true, which states that any triangle is a right triangle, if altitude is equal to the geometric mean of line segments formed by the altitude on the third side, is known as Geometric Mean Theorem (see the proof in the appendix). It gives rise to many interesting and nontrivial results that we will use in this and throughout the subsequent chapters.

Knowing how to construct the geometric mean of the two given segments is a very valuable tool in solving many construction problems.

PROBLEM 2.3.

Having two segments with length a and b, construct a segment with length $\sqrt{b^2 - a^2}$ (assuming that $b > a$).

SOLUTION.

Applying the formula for the difference of squares, $b^2 - a^2 = (a - b)(a + b)$, we can express $\sqrt{b^2 - a^2}$ as $\sqrt{b^2 - a^2} = \sqrt{(b - a)(b + a)}$. So, our problem again is reduced to constructing the geometrical mean of two segments, in this case, with lengths $b - a$ and $b + a$.

We can see that in light of the obtained results, we can construct many other segments. For instance, the segment of a $\sqrt{ab+cd}$ length can be constructed by making the following steps:

1. Construct segments with lengths $m = \sqrt{ab}$ and $n = \sqrt{cd}$.

2. Construct a segment with a length $\sqrt{m^2 + n^2}$.

The following problems 2.4 and 2.5 are good examples of real-life problems faced in allocating land spots of equal areas.

PROBLEM 2.4.

Draw a straight line parallel to one of the sides of the given triangle ABC such that it divides ABC into two parts that have equal areas.

<u>SOLUTION</u>.

Solution of every construction problem usually starts with analysis. Before we proceed to actual constructions, we need to contemplate the plan and justify our steps in the construction process.

Assume the problem is solved, and straight line $l \parallel AC$ is such that it divides the area of ABC in half. Denote M and N the points of intersection of l with AB and CB respectively. Draw $BK \perp AC$ and denote F the point of intersection of BK with MN. In triangles ABC and MBN, $\angle B$ is the common angle, $\angle A = \angle M$, and $\angle C = \angle N$ (as alternate interior angles by parallel lines). Therefore, these two triangles are similar, $\triangle ABC \sim \triangle MBN$.

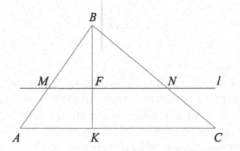

Theorem of Ratios of the Areas of Similar Polygons states that the ratio of the areas of the two similar polygons is equal to the squared ratio of their corresponding linear elements. Therefore, the ratio of the squares of two altitudes in these two similar triangles equals $\dfrac{1}{2}$, i.e., $\dfrac{BF^2}{BK^2} = \dfrac{S_{MBN}}{S_{ABC}} = \dfrac{1}{2}$. We obtain that $BF = \sqrt{\dfrac{1}{2}} \cdot BK$. Now, we can see that to construct the desired line l, we need to spot point F on the altitude BK such that

$BF = \sqrt{\dfrac{1}{2}} \cdot BK$, and then draw l at F parallel to AC. Since we already learned how to do constructions of irrational numbers, the last construction steps are manageable, and we hope that the readers will complete these steps independently.

It is noteworthy that in a similar manner we can divide triangle into two figures that have areas in any integer proportion.

PROBLEM 2.5.

Draw a straight line perpendicular to the side of a triangle in such a way that it divides this triangle into two parts that have equal areas.

<u>SOLUTION.</u>

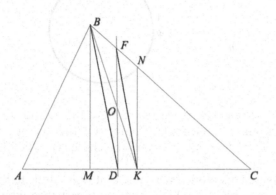

Let's analyze our problem and see if we can arrive at some good ideas on how to contemplate a plan for the construction.

Assume that FD is the desired segment, i.e. FD is such that $FD \perp AC$ and the areas of a quadrilateral $ABFD$ and right triangle FDC are equal. It implies that the area of FDC is half the area of our triangle ABC, $S_{FDC} = \frac{1}{2} S_{ABC}$.

We know that a median divides any triangle into two triangles of equal areas (see the appendix). Hence, after drawing median BK we get triangles ABK and CBK such that

$$S_{ABK} = S_{CBK} = \frac{1}{2} S_{ABC}.$$

Now, we observe that the areas of triangles FDC and CBK can be determined as

$$S_{FDC} = \frac{1}{2} FC \cdot CD \cdot \sin \angle C \text{ and } S_{CBK} = \frac{1}{2} BC \cdot KC \cdot \sin \angle C.$$

These areas are equal, therefore, we obtain that $FC \cdot CD = BC \cdot KC$, which can be written as $\frac{FC}{BC} = \frac{KC}{CD}$.

Draw now $BM \perp AC$. Since $FD \perp AC$, then $BM \parallel FD$, and verifying angles, we see that right triangles BMC and FDC are similar (all respective angles are congruent in both triangles). It follows that $\frac{CD}{MC} = \frac{FC}{BC}$. Recalling the derived earlier $\frac{FC}{BC} = \frac{KC}{CD}$, we get that $\frac{CD}{MC} = \frac{KC}{CD}$, from which $CD = \sqrt{MC \cdot KC}$. In other words, point D on AC has to be such that CD is the geometric mean of MC and KC. In the given triangle ABC, we know the length of MC and KC. So, our problem is reduced to constructing the geometric mean of the two segments with known lengths. Finding D, the last step will be to draw $FD \perp AC$. We leave all the construction steps for the readers to complete.

Working with intersections of circles and straight lines in construction problems, it is important to introduce the notion of an angle between a straight line and a circle. We will define an angle between a straight line and a circle as the angle between the straight line and the tangent line to the circle at their point of intersection. Since there are two such angles which are supplementary, we will always choose $0° \leq \varphi \leq 90°$.

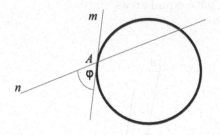

In a figure above, φ is the angle between the straight line n and the circle. It is the angle between n and the tangent m to the circle at point A, which is the point of intersection of the circle and n. If $\varphi = 90°$, the straight line and the circle are called orthogonal or perpendicular to each other.

We know that the tangent at a point on a circle is at right angle to the radius at this point. Therefore, the straight line orthogonal to the circle will pass through the center of the circle.

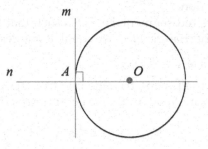

In a figure above, straight line n is orthogonal to the circle with center O because $n \perp m$ (m is the tangent at the point of intersection of n with the circle). Straight line n passes through O. This evolves immediately from the fact that one can draw the only one perpendicular to a line at a point on that line.

Next, we will define the angle between two circles. For two circles that intersect at a point A, the angle φ between them at A is the angle between their tangent lines at A. Similar to defining an angle between a straight line and a circle, since there are two such angles which are supplementary, we will always choose $0° \leq \varphi \leq 90°$.

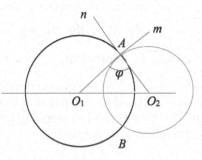

It is not hard to prove that a tangent to one of two orthogonal circles at the point of their intersection will necessarily pass through the center of the other circle.

Denote n and m the tangents to the orthogonal circles with centers O_1 and O_2, and let them intersect at point A. Then n is perpendicular to m, and the angle formed at A is 90°, $\angle A = \varphi = 90°$. Connecting O_1 with A we know that radius $O_1 A$ has to be perpendicular to the tangent passing through A, that is, $O_1 A \perp n$. There exists only one perpendicular to straight line n that can be dropped at point A laying on straight line n. We also know that m passing through A is perpendicular to n. Hence, we conclude that O_1 lies on m. In the same fashion we can prove that O_2 lies on n.

In light of our observations above, we can now easily tackle the following two construction problems 2.6 and 2.7.

PROBLEM 2.6.

Given two points A and B on the circle with center O_1 construct the circle passing through A and B that is orthogonal to the given circle with center O_1.

SOLUTION.

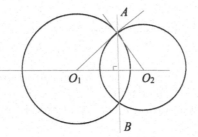

The center of the sought-after orthogonal circle can be found as the point of intersection of the perpendicular dropped from O_1 to AB and the tangent to the circle with the center O_1 passing through A. Such a tangent will be perpendicular to $O_1 A$ and, as proved earlier, will pass through O_2, the center of the desired orthogonal circle. After locating O_2, the last step will be to draw the circle with center O_2 and radius $O_2 A = O_2 B$.

The problem will have the unique solution when two perpendiculars will intersect. We suggest that readers investigate and prove that they will intersect when AB is a chord of the given circle different from a diameter. In case when points A and B are such that AB passes through O_1, i.e. when AB is the diameter of the circle, the problem has no solutions. It's noteworthy that in such a case AB is perpendicular to the circle with the center O_1.

PROBLEM 2.7.

For the given circle with center O_1 and point O_2 outside of the given circle, construct the circle with center O_2 which is orthogonal to the circle with center O_1.

SOLUTION.

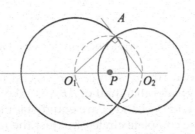

We connect points O_1 and O_2, and find P, the midpoint of the segment O_1O_2. Next, we draw the circle with center P and radius $r = \frac{1}{2}O_1O_2$. Designate the intersection of two circles point A. As the result of our constructions, $\angle O_1AO_2 = 90°$, because this inscribed angle in the circle with center P is subtended by diameter O_1O_2. Therefore, the circle with center O_2 and radius O_2A will be the sought-after orthogonal circle to the circle with center O_1.

The next construction is usually perceived as very simple and easy to do, but, in fact, it presents a real challenge to overcome.

PROBLEM 2.8.

Locate the mid-point of the given segment using compass only (no straightedge use is allowed).

SOLUTION.

Restrictions imposed on our allowable construction tools, to use a compass only, have a positive impact as well; they provide us with a hint on how to approach this construction problem.

Not being able to use a straightedge, we, in fact, cannot draw any straight lines. So, in all construction steps our goal will be to find the points as intersection of circles. The final destination, the mid-point D of our segment AB, should be determined as well as point of intersection of some circles.

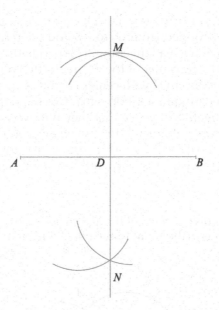

Let's first, discuss how this problem would be solved with two classic constructions tools, a straightedge and a compass.

We would need to draw two circles with centers at end points A and B and the same radius (its length should be greater than half of AB; to make it easy, it can be AB itself).

The points of their intersections M and N are equidistant from A and B (to get a clearer picture, we will not show the complete circles, but just the intersecting arcs that we need for the construction). Hence, the straight line passing through M and N is the locus of all such points equidistant from A and B. Draw MN intersecting AB at D, which is the desired mid-point of AB.

Solving, however, our original problem, we cannot use a straightedge and draw any straight lines, so we need to concentrate on finding some useful circles.

First, we will double our given segment AB, i.e. we will mark the point C on the straight line AB such that $AB = BC$. This can be done in the following steps:

1. Draw a circle with center at B and radius $r = AB$.

2. Mark on this circle points P, Q, and C such that $AP = PQ = QC = r$. To find P, we draw a circle with center at A and radius r and mark its intersection with the circle drawn in step 1. Next, draw a circle with center at P and radius r and mark its intersection Q with the circle drawn in step 1. Draw a circle with center at Q and find its intersection C with the circle drawn in step 1.

Since all three triangles APB, PBQ, and BQC are equilateral by construction, then $\angle ABC = 180°$. It implies that C is lying on AB, and $AB = BC$.

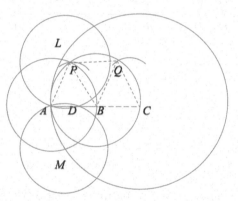

FIGURE 2.1

Now, we draw two circles, one with center at C and radius $CA = 2r$, and the second one with center at A and radius r, intersecting at L and M.

Finally, draw two circles with centers at L and M and radius r, each intersecting each other and segment AB at D. Let's prove that D is the mid-point of AB.

Notice that by construction, L and M are equidistant from AC, while D is equidistant from L and M. Hence, D must lie on AC. Consider two isosceles triangles ALD ($AL = LD$) and ACL ($CL = CA$). See Figure 2.2, the extract from Figure 2.1:

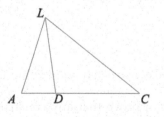

FIGURE 2.2

$\angle A$ is the common angle by the base of each of these isosceles triangles, hence the other angles are respectively congruent. Therefore, these triangles are similar, and it follows that $\frac{AD}{AL} = \frac{AL}{AC}$ or equivalently, $\frac{AD}{r} = \frac{r}{2r}$. From the last equality we obtain that $AD = \frac{1}{2}r = \frac{1}{2}AB$, which completes the proof.

This problem can be extended, and we can pose a more general question about how to divide a segment into n equal parts with the use of a compass only.

We repeat our constructions steps from the preceding problem to find C such that $AC = nAB$.

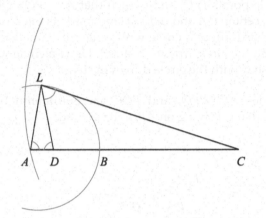

Then exactly in the same manner we find D on AC such that $AD = \frac{1}{n}AB$.

From similarity of isosceles triangles ALD and ACL it follows that $\frac{AD}{AL} = \frac{AL}{AC}$, which can be written as $\frac{AD}{r} = \frac{r}{AC}$, or equivalently, $AD \cdot AC = r^2$.

In geometry this result is related to a type of transformation of the Euclidean plane, called *inversion with respect to a circle*. Inversion transformation was first mentioned in a paper published in 1831 by German Jewish mathematician Ludwig Immanuel Magnus (1790–1861).

In the plane, the inverse of a point P with respect to a reference circle with center O and radius r is a point P', lying on the ray from O through P such that

$$OP \cdot OP' = r^2.$$

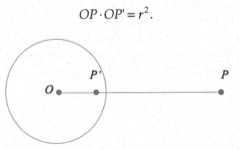

In fact, in our problem above we constructed point D, an image of C, in inversion with the circle with center A.

By utilizing inversion, one can prove that any construction problem solvable with a straightedge and compass can be solved with the use of a compass only. This statement is known as the *Mohr–Mascheroni theorem* after Danish mathematician Jorgen Mohr (1640–1697) and Italian mathematician Lorenzo Mascheroni (1750–1800). It should be understood that here we are talking about geometrical figures that contain no straight lines (it is impossible to draw a straight line without a straightedge). We know that a line is uniquely determined by any two distinct points on that line, so if two distinct points on a straight line are constructed, we understand that this line is clearly determined and "constructed" by us. The precise statement of the *Mohr–Mascheroni theorem* is:

Any Euclidean construction, insofar as the given and required elements are points (or circles), may be completed with the compass alone if it can be completed with both the compass and the straightedge together.

It is also noteworthy that a prominent Swiss mathematician Jacob Steiner (1796–1863) managed to prove the *Poncelet–Steiner theorem* that asserts that all problems of the second order can be solved with a straightedge alone without the use of a compass as soon as there is an additional restriction provided, namely, a circle and its center are given as one of the problem's conditions.

Next, it is natural to examine the following construction problem:

PROBLEM 2.9.

Given a circle with center O and a point P outside this circle, construct the image of P under inversion in the given circle.

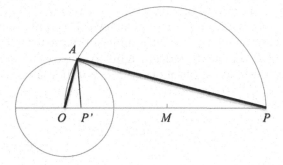

SOLUTION.

We have a circle with center O and radius r and point P that is outside of this circle. We need to find P' on OP such that $OP \cdot OP' = r^2$.

Construction steps:

1. Construct a circle with the center at M, the mid-point of OP, and radius $MO = MP$ intersecting our given circle at A.
2. Draw $AP' \perp OP$, $P' \in OP$.

We now have to prove that indeed $OP \cdot OP' = r^2$.

First, observe that $\angle OAP$ is a right angle because it is inscribed angle in a semicircle. We drew the altitude AP' to the hypotenuse in the right triangle OAP, so we can now apply an earlier-developed relationship in similar right triangles OAP ($\angle A = 90°$) and $OP'A$ ($\angle P' = 90°$) that $\dfrac{OP}{OA} = \dfrac{OA}{OP'}$. Recalling that $OA = r$, this is simplified to $OP \cdot OP' = r^2$, which is the desired result.

The readers can play with inversions by using dynamic geometry software such as Geometers Sketchpad, Cabri, or Cinderella. Using any of these you may construct the inverse image, P', of P under the inversion through the given circle. It should be interesting to investigate, for instance, the case of constructing an image of P when P is inside the given circle.

In some texts *inversion with respect to a circle* is called *symmetry in respect to a circle*. Point P' then is called symmetrical to point P in respect to a circle with center O, when P' lies on a ray OP and $OP \cdot OP' = r^2$ (where r is the radius of this circle).

It is important to recognize that inversion is not defined for the center of the given circle. This point does not have an image under inversion. As the chosen point gets closer to the center of inversion, its image gets farther away from the center. So, we can think of the center of inversion as transforming into a point at infinity P_∞ on the extended plane. We will talk about this in more detail in the next chapter discussing a geometric interpretation of two-dimensional hyperbolic geometry on the "unit circle" in the Poincaré Disk Model.

Ambitious readers might have guessed at this point that we did not come across symmetry in respect to a circle incidentally. We started our constructions explorations with geometrical mean of two segments and proceeded to utilizing it in other problems.

Step by step we were getting closer to the final objective of this chapter, getting acquainted with the notion of symmetry in respect to a circle, which in fact, is about constructing a geometrical mean of two segments. Rewriting the formula $OP \cdot OP' = r^2$ as $r = \sqrt{OP \cdot OP'}$, we see that the radius of inversion circle is the geometric mean of two segments OP and OP'.

The concluding problems in this chapter were selected deliberately to introduce this remarkable transformation of plain that flips the circle inside-out and is called inversion. It has many interesting and useful properties which we are going to investigate in the next chapter. Aside from providing alternative intriguing technique for solving many interesting problems, this topic can also provide some insights in unexpected geometry applications in real life. Not only are inversion applications useful, but they should enchant the readers.

3

Inversion and Its Applications

The mathematical facts worthy of being studied are those which, by their analogy with other facts, are capable of leading us to the knowledge of a physical law. They reveal the kinship between other facts, long known, but wrongly believed to be strangers to one another.

Jules Henri Poincaré

Inversion Properties

We will start with recalling the final constructions and the definition of inversion from the previous chapter: the inverse of a point P in a circle with center O and radius r is the point P' on the ray OP such that $OP \cdot OP' = r^2$. The transformation assigning to every point its inverse is called inversion.

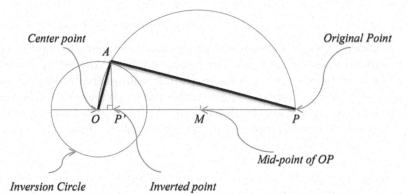

Inversion possesses several interesting properties:

Property 1. Points located inside the inversion circle are inverted into points outside of the circle, and vice versa. Also, If P' is the image of P, then P is the image of P'.

Property 2. Points on the inversion circle stay fixed, i.e. each point on the inversion circle inverts to itself.

Property 3. The angle between two lines is preserved under inversion, that is, inversion is a *conformal* transformation.

Property 4. The inverse image of a line passing through the center of inversion is the line itself.

DOI: 10.1201/9781003359500-3

It's not hard to see that the first four properties originated directly from the definition of inversion.

> *Property 5.* The inverse image of a circle not passing through the center of inversion is a circle not passing through the center of inversion.

> *Property 6.* The inverse image of a line not passing through the center of the inversion circle is a circle passing through the center of inversion.

Let's prove the last property (it looks the most intriguing out of all six).

PROOF.

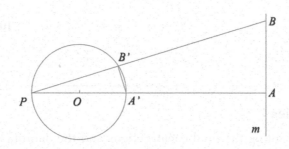

Let P be the center of the inversion circle (to make the diagram easier to read, we will not show this circle), and m is a straight line not passing through P. Draw $PA \perp m$, $A \in m$. Now, pick a random point B on m. Denote A' and B' the images of A and B in inversion with center P. Hence, by definition of inversion, $PA \cdot PA' = PB \cdot PB'$.

This can be rewritten as $\dfrac{PA}{PB} = \dfrac{PB'}{PA'}$. Considering triangles PAB and $PB'A'$ with the proportional corresponding sides and common angle P between them, the last equality allows us to conclude that they are similar. Since by construction, $\angle PAB = 90°$, then its respective angle in the similar triangle, $\angle PB'A' = 90°$ as well. It follows that B' lies on a circle with diameter PA' (angle $PB'A'$ is inscribed angle subtended by diameter PA'). Because point B was randomly selected on m, it implies that any point on m will be mapped into a point on a circle with diameter PA', which proves our assertion that the image of m is a circle passing through the center of inversion P.

The *converse statement* is true as well:

A circle passing through the center of inversion is transformed into a straight line not passing through the center of inversion.

Proof of this statement is similar to the proof of the direct statement; try to prove this as well as Property 5 yourself (or you can find the proof of property 5 in the appendix).

Very useful property of inversion is that it preserves angles. Therefore, if two circles or a circle and straight line are tangent at a point other than the center of inversion, then their inverse images are also tangent to each other; in a case when the point of tangency of two circles is the center of inversion then the inverse images are parallel straight lines (both are perpendicular to the straight line passing through the center of inversion and the centers of the tangent circles). Inversion is a nice alternative to traditional methods for solving geometrical problems, sometimes making certain problems much easier to solve. Since inversion transforms circles into straight lines and straight lines into circles, it will be the most useful whenever we are dealing with circles.

PROBLEM 3.1.

Prove that under inversion, the image of a circle orthogonal to the inversion circle is the same circle.

PROOF.

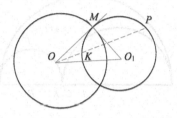

We consider two orthogonal circles, the circle of inversion with center O and the circle with center O_1. The goal is to prove that under the inversion in circle with center O, the circle with center O_1 will be inversed into itself.

Draw an arbitrary straight line through the center of inversion O which intersects the circle with center O_1 at two points K and P. Let's prove that K and P are the inverse of each other. In other words, we need to prove that $OP \cdot OK = r^2$, where r is the radius of the inversion circle.

Denote M the point of intersection of our two circles. Since circles are orthogonal, then OM is tangent to the circle with center O_1 (M is the point of tangency). By the *Tangent-secant theorem* (see the proof in the appendix), $OP \cdot OK = OM^2 = r^2$. This implies that indeed, K and P are the inverse to each other. In other words, we just proved that by drawing an arbitrary straight line through O that intersects the orthogonal circle in two points, we obtained that these points of intersection on the second circle are inverse to each other, i.e. the circle with center O_1 inverses to itself.

The next problem that we are going to solve is a very interesting problem that was offered in *Quantum* September/October 1992 issue, problem M63.

PROBLEM 3.2.

The diameter AB of a semicircle is arbitrarily divided into two parts, AC and CB, on which two other semicircles are constructed. Find the diameter of a circle inscribed in a curvilinear triangle formed by the three semicircles, given only the distance 1 from this circle's center to line AB.

SOLUTION.

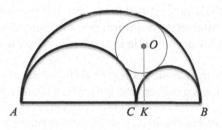

FIGURE 3.1

In our diagram in Figure 3.1, the red circle with center O touches three semicircles.

$OK \perp AB$ and it is given that $OK = 1$. The goal is to find the diameter of the red circle with center O. Let's invert all the given semicircles in inversion with a circle with center C and radius that is equal to the length of the tangent CF (F is the point of tangency) from C to the circle with center O (see Figure 3.2).

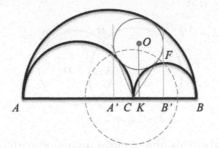

FIGURE 3.2

In other words, we are going to apply the inversion in a circle with center at the point of intersection of the two given semicircles and orthogonal to the given inscribed circle. Semicircles with diameters AC and BC will be inverted into the rays starting at points A' and B' on line AB (inverses of A and B) and perpendicular to AB (because both pass through the center of the inversion circle). Semicircle with diameter AB will be inverted into semicircle with diameter $A'B'$. Finally, our circle with center O will be inverted into a circle touching the two rays and the semicircle with diameter $A'B'$, as it is depicted in Figure 3.3. Recall now that we have chosen the inversion circle to be orthogonal to the given inscribed circle. Therefore, as it was proved in Problem 3.1, in such inversion our circle with center O will invert into itself.

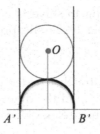

FIGURE 3.3

Note, that in our inversion, $A'B' = 2r = d$ (diameter of the given circle), and our circle touches the semicircle with diameter $A'B'$. Hence, the distance between their centers is also d (the distance from each center to the common point of tangency is r, so $r + r = d$). It implies that the sought-after diameter of the circle is 1, which completes the solution.

Next, we will turn to one of the classic geometrical theorems, *Ptolemy's theorem* (after Greek mathematician, astronomer, geographer, and astrologer Claudius Ptolemy (c.100–c.170 AD)) which states that

If a convex quadrilateral is inscribable in a circle then the product of the lengths of its diagonals is equal to the sum of the products of the lengths of the pairs of opposite sides.

Here is one of many possible solutions to this problem. It gives another nice example of utilizing inversion in problem-solving.

PROOF.

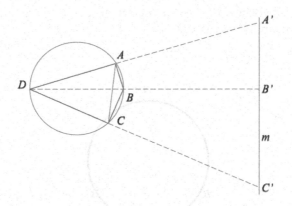

We are given cyclic quadrilateral $ABCD$ (it is inscribed in the circle). We need to prove that $AC \cdot DB = AB \cdot DC + DA \cdot BC$. Amazingly, this classic theorem can easily be interpreted by the fact that a circle passing through the center of inversion is transformed into a straight line perpendicular to a straight line passing through the center of inversion and the center of the circle.

Let's consider an inversion with D as the center of the inversion circle of radius r (we are not going to show this circle in a diagram to make it easily readable), and m that is a straight line image of the circumcircle of $ABCD$ in this inversion. Let's denote A', B', and C' the images of A, B, and C in inversion with center D. All three points A', B', and C' are collinear (they lie on m) with B' between A' and C', so

$$A'C' = A'B' + B'C'. \qquad (3.1)$$

By definition of inversion,

$$DA \cdot DA' = DB \cdot DB' = DC \cdot DC' = r^2. \qquad (3.2)$$

Checking similar triangles ADB and $B'DA'$, we see that $\dfrac{AB}{A'B'} = \dfrac{DA}{DB'} = \dfrac{DB}{DA'}$, from which $A'B' = \dfrac{AB \cdot DB'}{DA}$. Using (3.2), the last equality can be written as $A'B' = \dfrac{AB \cdot r^2}{DA \cdot DB}$. In a similar fashion we obtain that $B'C' = \dfrac{BC \cdot r^2}{DB \cdot DC}$ and $A'C' = \dfrac{AC \cdot r^2}{DA \cdot DC}$.

Substituting these expressions for $A'B'$, $B'C'$, and $A'C'$ into (3.1) gives

$$\frac{AC \cdot r^2}{DA \cdot DC} = \frac{AB \cdot r^2}{DA \cdot DB} + \frac{BC \cdot r^2}{DB \cdot DC}.$$

Multiplying both sides by $\dfrac{DA \cdot DB \cdot DC}{r^2}$ yields $AC \cdot DB = AB \cdot DC + DA \cdot BC$. This is the result we seek, and our proof is completed.

It merits mentioning that Ptolemy's theorem has many important corollaries, one of which is the famous Pythagorean Theorem. It would be a good exercise for the readers to prove it by making use of Ptolemy's theorem (or you can find such a proof in the appendix). Also, this classic theorem allows getting elegant and easy solutions to many seemingly difficult at first glance problems, for instance, the one examined below.

PROBLEM 3.3.

An equilateral triangle ABC is inscribed in a circle. Prove that for any randomly selected point D on this circumcircle $(D \neq A)$, $CD + BD = AD$.

SOLUTION.

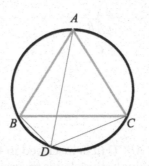

To simplify the following calculations, let's denote $AB = AC = BC = m$, $BD = x$, $CD = y$, and $AD = z$. Just by looking at the figure above, the natural desire is to shift the discussion from an inscribed triangle to a new figure, an inscribed convex quadrilateral $ABDC$, and try applying Ptolemy's theorem. AD and BC are the diagonals in $ABDC$. Thus, $BC \cdot AD = AB \cdot CD + AC \cdot BD$, or in our nominations, $mz = my + mx$. Canceling out $m(m \neq 0)$ on both sides of the last equality, we get the sought-after result, $z = y + x$, i.e., $CD + BD = AD$. Simple as that!

Many difficult construction problems become much more tractable when an inversion is applied. One such an example is the classic historical construction problem, *Apollonius's problem* (after ancient Greek geometer Apollonius of Perga (c.240 BC – c.190 BC). This problem, often called "the most famous geometry problem", is about constructing circles that are tangent to three given circles in a plane:

Given three circles in a plane (*Figure 3.4*), construct a circle (circles) tangent to each one of them (*Figure 3.5*).

If the circles are disjoint, not overlapping, and not contained within one another, then when the solution exists, generally, the construction should lead to eight distinct circles tangent to the three given circles.

We will examine the case when three circles do not intersect. This problem can be simplified to an easier version if we would have two of the three given circles tangent to each other. This can be achieved by increasing the radii of the circles.

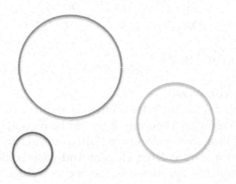

FIGURE 3.4

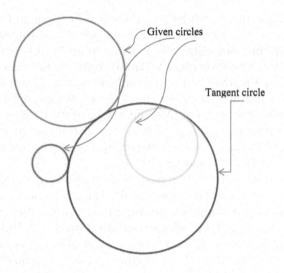

Given circles

Tangent circle

FIGURE 3.5

So, assume that we increased by the same length the radius of each of the given circles with centers O_1, O_2, and O_3 in such a way that circles with centers O_1 and O_2 became tangent at point O. Next, we will draw two tangent lines at O to the third of the given circles with center O_3. Denote points of tangency M and N. Now, we will apply an inversion in the circle with O as the center and radius $r = OM = ON$ (in order not to make our picture too overwhelming with various lines and circles, we will show not the full circle but just the dashed arc of that circle which passes through M and N). By the properties of a line tangent to a circle, we know that $O_3M \perp OM$. Hence, the third circle with center O_3 will map to itself in our inversion (see Problem 3.1). The other two circles with centers O_1 and O_2 will have as images two parallel lines m_1 and m_2 as two lines perpendicular to the same line (both circles pass through the inversion center and are tangent to each other, so they are inversed into straight lines perpendicular to the line passing through points O_1, O_2, and O). Our next step will be to draw the circle which is tangent to these two parallel lines and the circle with center O_3.

The center of such a circle must be equidistant from the two parallel lines m_1 and m_2. Hence, it lies on the line m parallel to m_1 and m_2 and is halfway between them.

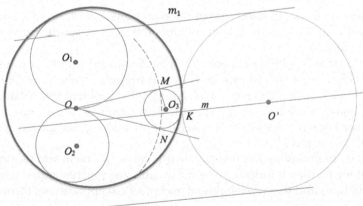

The radius of the circle is the distance between m and m_1 (or m and m_2, which is the same distance). Knowing this distance and having the point K of intersection of m with the circle with center O_3 as the point of tangency of our sought-after circle with the circle with the center O_3, it is easy to find its center O'. Finally, finding the reverse image of the green circle with center O' in our inversion will result in the sought-after red circle tangent to all three of our circles (because they are the reverse images of the constructed parallel lines and the circle with the center at O_3 – the image of itself). Recalling that at the very beginning of our constructions we increased the radii of the given circles by the same length, we would need to make an extra step reducing the radius of each circle and the circle tangent to each of them by that length.

We did not explain in detail the actual basic constructions in all our intermediary steps. It should be a good exercise for the readers to thoroughly perform every step in the construction process and its justification, determine how unique the solution is, and review alternative solutions depending on various configurations of the three given circles.

This classic problem has many different solutions; applying inversion properties is only one of them. Remarkably, even algebraic methods were implemented, which transform this geometric problem into algebraic equations. In 1643 the prominent French philosopher and mathematician René Descartes (1596–1650) asserted that for every four mutually tangent circles the radii of the circles satisfy a certain quadratic equation. This statement is known as *Descartes' Theorem,* and it provides an algebraic solution of a specific Apollonius's problem case for three given mutually tangent circles. By solving a quadratic equation that relates the radii of these four circles, we can construct a sought-after circle tangent to three given, mutually tangent circles. In a general case, considering three given circles (not necessarily mutually tangent) and the seek-after fourth circle tangent to all of them in the Cartesian coordinate system in the two-dimensional plane, we know that their positions can be determined in terms of the (x, y) coordinates of their centers. Denoting coordinates of the centers of three given circles (x_1, y_1), (x_2, y_2), (x_3, y_3), and coordinates of the center of a desired solution circle (x_4, y_4) and introducing their radii respectively as r_1, r_2, r_3, and r_4, we relate the requirement that a solution circle must be tangent to each of the three given circles by the following system of three quadratic equations for x_4, y_4, and r_4:

$$\begin{cases} (x_4 - x_1)^2 + (y_4 - y_1)^2 = (r_4 - s_1 r_1)^2, \\ (x_4 - x_2)^2 + (y_4 - y_2)^2 = (r_4 - s_2 r_2)^2, \\ (x_4 - x_3)^2 + (y_4 - y_3)^2 = (r_4 - s_3 r_3)^2. \end{cases}$$

Here the numbers on the right-hand side of the equations $s_1 = \pm 1$, $s_2 = \pm 1$, and $s_3 = \pm 1$ indicate whether the desired solution circle should touch the corresponding given circles internally ($s = 1$) or externally ($s = -1$).

It is noteworthy that Apollonius's problem provides an incredible example of practical applications of classic geometrical problems in real life when solving certain problems of mathematical physics, optics, and electricity. Apollonius' problem formulated as the problem of locating one or more points such that the differences of its distances to three given points equal three known values finds its useful application in navigation systems such as *Decca Navigator System* and *LORAN*.

We suggest that ambitious readers investigate this classic problem and its numerous applications and history further. It indeed presents an amazing journey into the history of geometrical constructions and their development and practical applications through the ages.

Mechanical Engineering Applications of Inversion

Remarkably, the pure geometrical topic of the inversion in a circle is related to many mechanical engineering applications. In fact, the theory of linearizing mechanisms that convert circular motion to linear motion is based on the properties of inversion. Such a change of motion from circular to linear occurs in many different mechanical settings. The study of the devices capable of such a conversion was particularly important in the development of steam engines.

The first true planar straight-line mechanism called *Peaucellier–Lipkin inversor* or *Peaucellier–Lipkin linkage* after its inventors the French engineer Charles-Nicolas Peaucellier (1832–1913) and Lithuanian mathematician and inventor Yom Tov Lipman Lipkin (1846–1876) is directly related to the inversive geometry. Peaucellier constructed the inversor in 1864. Lipkin independently discovered the linkage (it is also known as *Lipkin parallelogram*) in 1871.

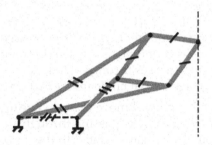

It was the first planar linkage capable of transforming rotary motion into perfect straight-line motion, and vice versa. The device is described as the following:

Six bars are hooked together with flexible joints into one "geometrical construction" - four bars of the same length form a rhombus and two other bars, also of the same length but longer than the rhombus's sides, are attached to opposite vertices of the rhombus.

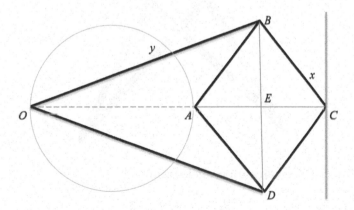

In the figure above, $AB = BC = CD = DA$ and $OB = OD$. If point O is fixed and point A moves along a curve, then point C moves along the image of that curve under inversion with respect to a circle centered at O and radius $r = \sqrt{OB^2 - AB^2}$. To prove this assertion, we need to prove that regardless of Peaucellier–Lipkin inversor's position, the product of OA and OC is a constant number r such that $r^2 = OB^2 - AB^2$.

To simplify our calculations, we denote $AB = BC = CD = DA = x$ and $OB = OD = y$. OE is the median and the altitude in isosceles triangle BOD, where E is the point of intersection of the diagonals of the rhombus $ABCD$. Clearly, the vertices A and C of the isosceles triangles BAD and BCD will lie on the straight line through O and E (since the diagonals of the rhombus AC and BD are perpendicular and bisect each other at E).

Consider $\triangle BEO$, ($\angle E = 90°$). Applying the Pythagorean Theorem yields

$$BE^2 = OB^2 - OE^2 = y^2 - OE^2. \tag{3.3}$$

Applying the Pythagorean Theorem to right triangle $\triangle BEC$, ($\angle E = 90°$) we have

$$BE^2 = BC^2 - CE^2 = x^2 - CE^2. \tag{3.4}$$

Equating (3.4) and (3.3) gives $y^2 - OE^2 = x^2 - CE^2$, from which, observing that $AE = CE$, we get that $y^2 - x^2 = OE^2 - CE^2 = (OE - CE)(OE + CE) = OA \cdot OC$. We obtained that the product of OA and OC is indeed a constant number $\left(y^2 - x^2\right)$, expressed in terms of the lengths of our bars forming the inversor. So, if point O is fixed and if point A is constrained to move along a circle (green circle) which passes through O, then point C will have to move along a straight line (red line), the image of that circle under inversion with center O and radius $r = \sqrt{y^2 - x^2}$. On the other hand, if point A moves along a line (not passing through O), then point C would have to move along a circle (passing through O).

The linearizing articulation mechanisms related to the inventions of English mechanical engineer James Watt (1736–1819) and Russian mathematician and engineer Pafnuty Chebyshev (1821–1894) are based on the same principles.

Another example of an inversor that converts rotary motion to straight line motion is presented by Hart's inversor (after English mathematician and engineer Harry Hart (1848–1920)). This invention was published in 1874.

Hart's inversor is based on an antiparallelogram.

Antiparallelogram is a quadrilateral in which its two pairs of opposite sides are equal and two of them in the longer pair intersect each other. In the antiparallelogram $ACBD$ in the diagram above, $AD = BC$, $AC = BD$, and AC intersects BD.

Antiparallelogram possesses an important property that the product of its diagonals, $DC \cdot AB$, is a constant value. But before we proceed to a direct proof of this assertion, we need to preface the proof with several observations proving first the very important fact that in an antiparallelogram, $AB \parallel DC$.

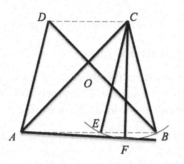

Triangles ADB and BCA are congruent by three sides ($AC = BD$, $AD = BC$ as given, and AB is the common side). Therefore, their respective angles are congruent as well, namely, $\angle ABD = \angle BAC$, and denoting O the point of intersection of AC and BD, it follows that triangle AOB is isosceles. In a similar fashion, we can prove that triangle DOC is isosceles as well. But since these two isosceles triangles have equal vertical angles at vertex O, it implies that the angles by their bases are equal as well (each can be found as one-half of the difference between $180°$ and angle O). The respective angles by each base are alternate interior angles by DC and AB, hence, it immediately follows that indeed, as we wished to prove, $AB \parallel DC$. Now, our preliminary work has been done and we are set for the proof of our assertion that the product of the diagonals of $ACBD$, $DC \cdot AB$, is a constant value.

We start with drawing $CE \parallel DA$, $E \in AB$, which makes $ADCE$ a parallelogram (it has two pairs of parallel opposite sides, as we just proved that also $AB \parallel DC$). Next, we draw a circle with center C and radius $CE = CB$ (in our parallelogram $ADCE$, $CE = DA$ and it is given that $DA = BC$, so, by transitivity, $CE = CB$) and draw the line AF tangent to this circle. By the tangent-secant theorem, $AF^2 = AE \cdot AB = DC \cdot AB$. Observe that since AF is tangent to our circle, then drawing radius CF, we get that $CF \perp AF$, and applying the Pythagorean Theorem to right triangle AFC, we get $AF^2 = AC^2 - CF^2$. Comparing the two expressions for AF^2 gives $DC \cdot AB = AC^2 - CF^2$. Since AC and $CF = CE = AD$ are fixed as the edges in antiparallelogram linkage, their lengths are constant numbers known to us, and we arrive at the desired result; indeed, the product of the diagonals of antiparallelogram is a constant number.

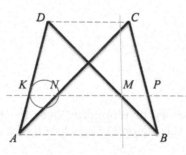

Next, we fix a point on one of the sides, let's say, K on AD, and draw the straight line at K parallel to DC and AB intersecting at N, M, and P the sides AC, BD, and BC respectively. We will prove that the product of the distances from the fixed point K to the intersections of our parallel line with AC and BD remains the same for all positions of the antiparallelogram, that is, $KN \cdot KM$ is some constant number.

Considering angles, we get two pairs of similar triangles, $\Delta AKN \sim \Delta ADC$ and $\Delta ADB \sim \Delta KDM$. It follows that $\dfrac{KN}{DC} = \dfrac{AN}{AC}$ and $\dfrac{KM}{AB} = \dfrac{DM}{DB}$. Multiplying these equalities and recalling that $AC = BD$, we can express the product in question as $KN \cdot KM = DC \cdot AB \cdot \dfrac{AN \cdot DM}{AC^2}$. Clearly, the ratio $\dfrac{AN \cdot DM}{AC^2}$ is some constant number because all its terms are constant numbers for a fixed position K on AD, and the product $DC \cdot AB$ is a constant number, as it was proved earlier. Therefore, $KN \cdot KM$ is some constant value, as we wanted to prove. This implies that N and M are mutually inverse under inversion with center K and radius $r = \sqrt{DC \cdot AB \cdot \dfrac{AN \cdot DM}{AC^2}}$. This means that if K is fixed and N is moving along a curve, M will be moving along the inverse image of the curve.

If N moves along the red circle passing through K, then M will move along the green straight line.

Instead of $KN \cdot KM$ we can consider any other equal product in our configuration and use exactly analogous reasoning to the above. Therefore, for Hart's inversor, taking any of the above four points as the center of inversion, we can move the second point along the circle passing through the first point, enabling the third point moving along the straight line.

Non-Euclidean Geometries

For two millennia, Euclid's geometry was held as the only sort of geometry that had been conceived and regarded as absolute geometry. But if the first four postulates deem self-evident than the fifth Euclid's postulate did not look obvious, and it was not perceived as natural like the other four. Euclid formulated it in a very "cautious" manner and perhaps included it on his list only when he gave up the attempts to prove it and was not able to proceed without it. In fact, in the modern world it is better known not as it was originally stated by Euclid, but as the statement formulated by Scottish mathematician John Playfair (1748–1819):

> *In a plane, given a line and a point not on it, at most one line parallel to the given line can be drawn through the point.*

Mathematicians had a long history of wrangling over the fifth postulate, and lots of attempts were made to prove it relying on Euclid's first four postulates. All such attempts failed.

The Russian mathematician Nikolai Lobachevsky (1792–1856) developed geometry where the "parallel postulate" does not hold. Instead, he suggested allowing more than one line to be defined as non-intersecting a given line through a point not on that line. In his geometry, all fundamental premises are the same as in Euclidean geometry, except for the axiom of parallelism. He presented his ideas for the first time in 1826 to the session of the Kazan University department of physics and mathematics. In 1829 Lobachevsky described his revolutionary ideas in "On the Origin of Geometry" ("О Началах геометрии" – in Russian) which was published by the *Kazan Messenger*. The Hungarian mathematician János Bolyai (1802–1860) independently developed similar ideas. In 1831 he described new geometry in the appendix of the book published by his father Farkas Bolyai. From the correspondence between Farkas Bolyai and Carl Gauss, we know that Gauss admitted that similar ideas occupied him for more than thirty or thirty-five years. However, he had never published his findings perhaps because he viewed the non-Euclidean geometry as too revolutionary to be revealed at that time. Indeed, those ideas were deemed so unbelievable and unconventional. Their reception by the mathematical community was not very enthusiastic (to say at least!) at the time. Even though Lobachevsky declared his new geometry to be self-consistent, he did not have a model to prove its consistency. His Russian mathematicians-contemporaries simply did not understand his ideas. St. Petersburg Academy of Sciences rejected his publication in *Kazan Messenger*, and it did not draw proper attention in mathematicians' circles. After his dismissal from his position as rector of Kazan University due to his deteriorating health, he died in poverty at age of 63.

Not less tragic was the fate of "… a genius of the first order" (that is how Carl Gauss regarded János Bolyai in a letter to his friend after reading the Appendix in Farkas Bolyai's book). This Appendix was János Bolyai's only published work, even though he left more than 20,000 pages of mathematical manuscripts when he died in obscurity at age of 58. It is tragic and very sad that Nikolai Lobachevsky and János Bolyai did not get due recognition for their remarkable work during their lifetime.

To visualize the new non-Euclidean geometry and to accept it as logically consistent, a model was needed which would allow the study of its properties. Such a model should provide an interpretation of the Euclidean plane, points, and straight lines.

In conclusion of our discussion of inversion and its applications we should mention the very important role that inversion plays in the Poincaré Disk Model which is used to prove that Bolyai-Lobachevskian geometry, or how it is often called, hyperbolic geometry, is consistent if and only if Euclidean geometry is consistent. In our days, this remarkable model allows us to perform the constructions via Sketchpad and be visualized within the Euclidean plane. This model is named after the great French mathematician, Jules Henri Poincaré (1854–1912), but it was Italian mathematician Eugenio Beltrami (1835–1900), who first proposed in 1868 that non-Euclidean geometry could be realized on a surface of constant negative curvature, a pseudosphere.

Beltrami attempted to show that two-dimensional non-Euclidean geometry is as valid as the Euclidean geometry of the space. His findings indicated that Euclid's parallel postulate could not be derived from the other axioms of Euclidean geometry. Today Beltrami's work is recognized as very important in the development of non-Euclidean geometry; however, such recognition was not fully credited to him at the time. Only 14 years later the Beltrami's rediscovered model was proposed by Henri Poincaré and obtained a new life.

Henri Poincaré studied two models of hyperbolic geometry, one based on the open unit disk, and the other on the upper half-plane.

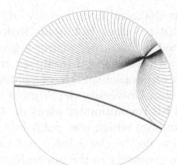

In the Poincaré Disk Model, we consider a unit circle (any circle in the Euclidean plane with a radius of one) and define the points of the hyperbolic plane as points in the interior of this circle. The points on the circumference of our unit circle are not defined as points that lie in the hyperbolic plane. Such points are called *vanishing points, ideal points,* or *points at infinity.* The straight lines of the hyperbolic plane are defined as arcs of circles that are contained in the interior of the unit circle and are orthogonal to the unit circle, and all the diameters of the unit circle (without endpoints, as those are the points at infinity and do not belong to the hyperbolic plane).

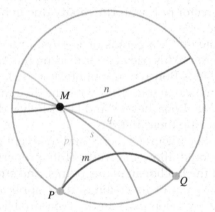

In the figure above, there are two lines p and q passing through M and "parallel" (that's how Nikolai Lobachevsky called these two lines) to line m. Each of these two lines "touches" m at points P and Q at infinity. There are infinitely many other lines through M that do not intersect m. Line n is one of such lines. There are also infinitely many other lines through M that intersect m. Line s is one of such lines.

    To delve into hyperbolic geometry, one has to refer to the theory of inversion, which is essential for understanding the Poincaré model of non-Euclidean geometry. The described model is simply the image of the Euclidean plane after inversion. To be able to do the compass and straightedge constructions of a hyperbolic line through two points of the hyperbolic plane one has to be familiar with inversion and its properties! Knowledge of inversion allows actually constructing specific isometric transformations of the Poincaré Disk. The detailed discussion of such constructions and properties of hyperbolic geometry is beyond the scope of our book. The revolutionary discovery of non-Euclidean geometry is one of the greatest accomplishments in the history of science. It had a similar sort of effect on the world view of humanity as the great discovery of the heliocentric system of Copernicus. No wonder that Nikolai Lobachevsky was called the Copernicus of Geometry! It completely changed the way mathematicians and scientists view the Universe, and it originated the search for other "new geometries". Lobachevsky said, "… there cannot be any contradiction in our minds when we suppose that some forces in nature follow one Geometry and others follow their own special Geometry." At the end of the 19th century Henry Poincaré and Felix Klein (1849–1925) (German mathematician known for his work with non-Euclidean geometry, complex analysis, and on associations between geometry and group theory) established that Bolyai-Lobachevskian geometry is closely connected to the theory of functions of a complex variable and to number theory. Amazingly, as we know now, non-Euclidean geometries are broadly applied in physics and other branches of science. For instance, the space of relativistic particles, i.e. particles moving at velocities close to the speed of light, is governed by Bolyai-Lobachevsky's geometry. The velocity space of the special theory of relativity is a Lobachevskii space. American mathematician William Thurston (1946–2012) discovered important connections of hyperbolic geometry with the topology of 3-manifolds. So, those non-Euclidean geometries are real. Even though we feel that the space where we live is governed by Euclidean geometry rather than non-Euclidean geometries, these geometries may be applied to other spaces that appear in mathematical and physical theories. As Henri Poincaré stated, "One geometry cannot be more true than another; it can only be more convenient"!

We hope that the reader's exposure to the discussed fascinating topics will entice further exploration.

4

Using Geometry for Algebra. Classic Mean Averages' Geometrical Interpretations

Equations are just the boring part of mathematics.

I attempt to see things in terms of geometry.

Stephen Hawking

When solving many algebraic problems, it is worthwhile to explore, if possible, an alternative geometrical approach. One may get unexpected vivid and enlightening interpretations of the given conditions leading to elegant and easy solutions. In some instances, the simple idea to give a problem a geometric interpretation allows us to convert a problem into a new one that is easier to solve rather than do it directly.

PROBLEM 4.1.
Prove that the sum of consecutive odd numbers starting from 1 is a perfect square.

SOLUTION.
Very interesting geometrical solution to this presumably algebraic problem was presented in the Pythagorean School in Ancient Greece.

Greek mathematicians envisioned 1 (a unit) as a square, and all next odd numbers as geometric gnomons, figures of a Γ- form, consisting of an odd number of squares (units):

$$1+3 = 4 = 2^2,$$
$$1+3+5 = 4+5 = 3^2,$$
$$1+3+5+7 = 9+7 = 4^2.$$

They imagined these squares as built up of gnomons added to unity. For example, they saw that 1 + 3, 1 + 3 + 5, 1 + 3 + 5 + 7, and so on, are squares and that the odd numbers in the figure below were related to the geometric gnomon. Such numbers were, therefore, themselves called gnomons.

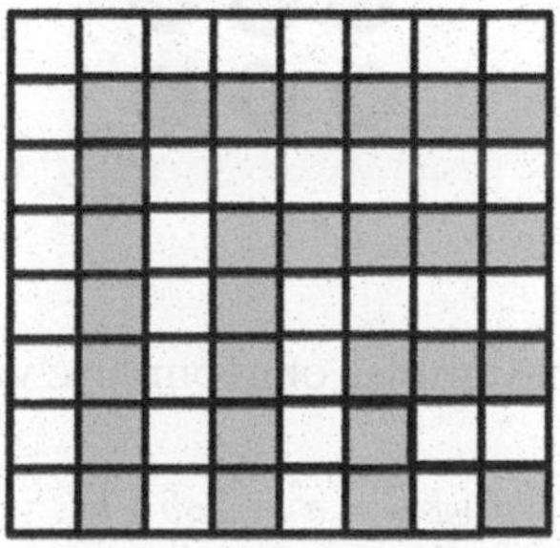

DOI: 10.1201/9781003359500-4

This problem has a simple algebraic solution as well. The sequence of all odd numbers starting from 1 is an arithmetic progression with the common difference of 2:

$$1, 3, 5, 7, 9, \ldots, (2n+1).$$

The sum of the members of a finite arithmetic progression can be calculated by taking the number n of terms being added, multiplying by the sum of the first and the last number in the progression, and dividing by 2:

$$S_n = \frac{a_1 + a_n}{2} \cdot n.$$

Considering $(n+1)$ terms in our sequence, we can apply the general formula to calculate the sum of the first $(n+1)$ numbers:

$$S_{n+1} = \frac{1+(2n+1)}{2} \cdot (n+1) = \frac{2n+2}{2} \cdot (n+1) = (n+1) \cdot (n+1) = (n+1)^2.$$

Mathematicians in Pythagorean School used application of geometric gnomons to solve the following problem as well (try to prove this yourself before reading the solution):

PROBLEM 4.2.
Prove that any odd number, other than 1, can be expressed as the difference of two squares.

<u>SOLUTION</u>.
Indeed, referring to the figure in the previous problem, we see that if one cuts a gnomon representing an odd number from a square, the remaining figure will be a square, i.e., $(2n+1)+n^2 = (n+1)^2$ or equivalently, $2n+1 = (n+1)^2 - n^2$, and our assertion is proved.

We can also show a vivid geometrical illustration of the "reversal" sum of odd numbers $1+3+5+\cdots+(2n-1)+\cdots+5+3+1 = (n-1)^2 + n^2$.

For $n = 4$, we see that indeed, the sum of $1+3+5+7+5+3+1$ circles in the left figure equals to the sum of $9 = 3^2$ green circles and $16 = 4^2$ white circles shown separately in the two right figures.

A few more vivid and appealing examples of geometric views of algebraic formulas:

$$(a+b)^2 = a^2 + 2ab + b^2.$$

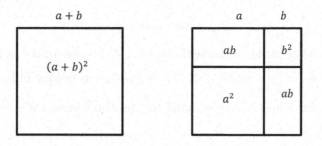

$$1^3 + 2^3 + 3^3 + 4^3 + \cdots + n^3 = (1+2+3+4+\cdots+n)^2$$

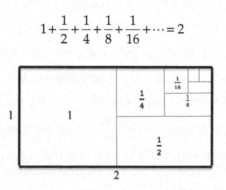

To grasp the above geometrical interpretation, we can rewrite the sum of cubes of natural numbers as $1 \cdot 1^2 + 2 \cdot 2^2 + 3 \cdot 3^2 + 4 \cdot 4^2 + 5 \cdot 5^2 + 6 \cdot 6^2 + \cdots + n \cdot n^2$ and observe that even though the even squares overlap forming yellow squares, there are also uncovered gaps in red squares negating our overlaps (because each red square is congruent to the respective yellow square!).

Next, let's look at the geometric interpretation of the sum of infinite geometric sequence with the ratio $q = \frac{1}{2}$:

$$1 + \frac{1}{2} + \frac{1}{4} + \frac{1}{8} + \frac{1}{16} + \cdots = 2$$

We consider a rectangle of area 2 and cut it into two rectangles of equal areas, each of 1. Then we cut one of those two rectangles again in half and continue this process getting rectangles of areas $\frac{1}{2}, \frac{1}{4}, \frac{1}{8}, \frac{1}{16}, \ldots, \frac{1}{2^n}, \ldots$ The sum of the areas of all these small rectangles comprising the big one is equal to its area, that is, to 2.

Hence, $1+\dfrac{1}{2}+\dfrac{1}{4}+\dfrac{1}{8}+\dfrac{1}{16}+\cdots=2$, as we wanted to prove.

The readers should be well familiar with the formula for the sum of an infinite geometric sequence $1+q+q^2+\cdots=\dfrac{1}{1-q}$ when $|q|<1$. The suggested technique utilizing the geometric interpretation can be found handy proving the general formula above for $q=\dfrac{1}{n}$ for any natural $n\neq0$: $S=1+\dfrac{1}{n}+\dfrac{1}{n^2}+\cdots=\dfrac{1}{1-\dfrac{1}{n}}=\dfrac{n}{n-1}$ (try reproduce this proof on your own starting with a rectangle of area n and cutting it into n equal rectangles of area 1). Also, this technique should be appealing in finding some other infinite sums, as for example, it is shown in the following Problem 4.3.

PROBLEM 4.3.

Prove that $1+\dfrac{1}{2!}+\dfrac{2}{3!}+\dfrac{3}{4!}+\cdots+\dfrac{n}{(n+1)!}+\cdots=2.$

SOLUTION.

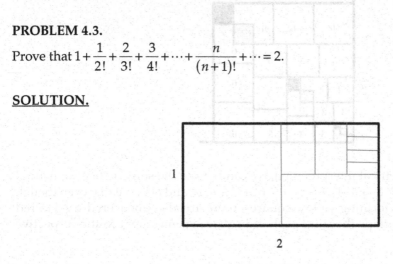

Similarly to the previous problem, we'll cut the rectangle of area 2 into two equal rectangles and then cut one of them into two rectangles of the same area of $\dfrac{1}{2}$. In the next step we cut one of the last rectangles into three congruent rectangles, each with the area of $\dfrac{1}{2}:3=\dfrac{1}{6}=\dfrac{1}{2\cdot3}=\dfrac{1}{3!}$. One of these rectangles of area $\dfrac{1}{6}$ we will cut into four congruent rectangles, then one of such rectangles is cut into five congruent rectangles, and so on. Continuing this process, we obtain one rectangle of area 1, one rectangle of area $\dfrac{1}{2!}$, two rectangles of area $\dfrac{1}{3!}$, three rectangles of area $\dfrac{1}{4!}$, ..., n rectangles of area $\dfrac{1}{(n+1)!}$, and so on.

Adding all the areas of small rectangles comprising the big one, we get their sum to be equal to 2, the area of the big rectangle, i.e. the desired equality:

$$1+\dfrac{1}{2!}+\dfrac{2}{3!}+\dfrac{3}{4!}+\cdots+\dfrac{n}{(n+1)!}+\cdots=2.$$

One of the subsequent chapters will be devoted to solving inequalities and different techniques for proving algebraic inequalities. Ambitious readers should come back to the following Problem 4.4 after reading that chapter and try some other approach for solving it. Here we suggest examining the geometrical illustration of the given inequality which enables us to solve the problem fairly quickly.

PROBLEM 4.4.

There are given positive numbers a, b, and c ($a > 0$, $b > 0$, $c > 0$). Prove that

$$\sqrt{a^2 - ab + b^2} + \sqrt{b^2 - bc + c^2} > \sqrt{a^2 + ac + c^2}.$$

SOLUTION.

Let's consider two triangles AOB and BOC with the common side BO such that $AO = a$, $BO = b$, $CO = c$, and $\angle AOB = \angle BOC = 60°$.

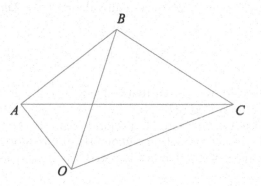

We will apply the Law of cosines three times, first to triangle AOB, second to triangle COB, and finally, to triangle AOC (note that $\angle AOC = \angle AOB + \angle BOC = 120°$).

Recall that $\cos 60° = \frac{1}{2}$ and $\cos 120° = -\frac{1}{2}$. Therefore,

in triangle AOB we have $AB = \sqrt{a^2 + b^2 - 2ab \cdot \cos 60°} = \sqrt{a^2 - ab + b^2}$;

in triangle COB we have $BC = \sqrt{b^2 + c^2 - 2bc \cdot \cos 60°} = \sqrt{b^2 - bc + c^2}$;

in triangle AOC we have $AC = \sqrt{a^2 + c^2 - 2ac \cdot \cos 120°} = \sqrt{a^2 + ac + c^2}$.

As you can see now, we expressed three sides of the newly formed triangle ABC in terms of a, b, and c. The Triangle Inequality theorem states that the sum of any two sides of a triangle is greater than the third side. Therefore, in triangle ABC, $AB + BC > AC$. This implies that $\sqrt{a^2 - ab + b^2} + \sqrt{b^2 - bc + c^2} > \sqrt{a^2 + ac + c^2}$, which is the sought-after result. Our inequality is proved.

There are many problems involving algebraic inequalities in which adding geometrical interpretation facilitates the solution or opens a new nice and unexpected approach.

Since inequalities occupy important place in our explorations throughout the book and we will refer to them on many occasions in our subsequent chapters, we will concentrate in the rest of this chapter on classic algebraic means which appear repeatedly in algebra problem books, and their geometrical interpretations.

We will start with inequality known as *The Arithmetic Mean – Geometric Mean (AM – GM) inequality*. We will consider its simplest case for two numbers.

The Arithmetic Mean – Geometric Mean (AM – GM) Inequality

The arithmetic mean of two nonnegative real numbers is greater than or equal to their geometric mean:

$$\frac{a+b}{2} \geq \sqrt{ab}, \text{ where } a \geq 0, b \geq 0$$

with equality if and only if $a = b$.

This well-known inequality is easy to prove algebraically using the obvious fact that the square of any number is a nonnegative number, $(a-b)^2 \geq 0$.

Squaring the difference on left-hand side gives $a^2 - 2ab + b^2 \geq 0$. Adding and subtracting $4ab$, we get $a^2 - 2ab + b^2 + 4ab - 4ab \geq 0$, or equivalently, $(a^2 + 2ab + b^2) - 4ab \geq 0$. The last expression is simplified to

$$(a+b)^2 \geq 4ab.$$

Recalling that $a \geq 0$ and $b \geq 0$, we obtain that indeed, $\frac{a+b}{2} \geq \sqrt{ab}$, with equality attained only when $a = b$, as it was required to be proved.

There are few appealing geometrical interpretations of this property. The idea underlying the first one is related to constructing a geometrical mean of two arbitrary segments similar to how we did this construction in Chapter 2 for a unitary segment and some segment of length x.

Let's recall the construction steps.

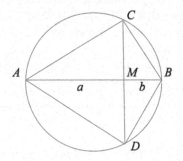

Consider two segments with length a and b each and draw the segment with length $(a+b)$, $AM = a$, $MB = b$, $AB = a + b$. Next, we construct the circle with center at mid-point of AB and with radius $r = \frac{a+b}{2}$, and at M draw the perpendicular to AB, meeting our circle at points C and D. Then $\angle ACB = \angle ADB = 90°$ as inscribed angles subtended by dimeter AB, and by Geometric mean theorem, $CM = \sqrt{AM \cdot BM} = DM$ or substituting for a and b, $CM = DM = \sqrt{a \cdot b}$.

Hence, $CD = 2CM = 2\sqrt{a \cdot b}$. Being a chord of the considered circle, clearly, CD in length does not exceed the diameter of the circle, i.e. $CD \leq AB$, or in our nominations, we obtain that $2\sqrt{a \cdot b} \leq a + b$. This can be written as $\frac{a+b}{2} \geq \sqrt{a \cdot b}$, with the equality possible only when CD is a diameter, or, in other words, when $a = b$.

The full inequality is an extension of this idea to n dimensions. This inequality is also known as the *Cauchy's Inequality* after the prominent French mathematician Augustin-Louis Cauchy (1789–1857):

The arithmetic mean of any n nonnegative real numbers is greater than or equal to their geometric mean. The two means are equal if and only if all the numbers are equal:

$$\frac{a_1 + a_2 + \cdots + a_n}{n} \geq \sqrt[n]{a_1 \cdot a_2 \cdot \ldots \cdot a_n}.$$

This can be proved by mathematical induction. Try to prove this yourself or you can find a proof in a standard algebra textbook.

The other well-known inequalities that permeate many branches of mathematics concern the relationships connecting AM – GM inequality with so-called *harmonic mean* and *quadratic mean*.

The *harmonic mean H* of the positive numbers $a_1, a_2, \ldots, a_n$ is defined as

$$H = \frac{n}{\dfrac{1}{a_1} + \dfrac{1}{a_2} + \ldots + \dfrac{1}{a_n}}.$$

The Quadratic mean (or *Root mean square*, it is usually abbreviated *RMS*), is defined as

$$RMS = \sqrt{\frac{a_1^2 + a_2^2 + \ldots + a_n^2}{n}}.$$

The following inequalities are true for any positive numbers a_k ($k = 1, 2, \ldots, n$) and any natural number n ($n \neq 0$):

$$\frac{n}{\dfrac{1}{a_1} + \dfrac{1}{a_2} + \ldots + \dfrac{1}{a_n}} \leq \sqrt[n]{a_1 \cdot a_2 \cdot \ldots \cdot a_n} \leq \frac{a_1 + a_2 + \ldots + a_n}{n} \leq \sqrt{\frac{a_1^2 + a_2^2 + \ldots + a_n^2}{n}}.$$

The discussion of algebraic means gives us great opportunity to consider various geometrical interpretations to justify our algebraic inequalities. The idea to apply geometrical approach to algebraic problems is mutually beneficial. On one hand, it introduces a different view at algebraic problems helping "visualizing" them and making certain problems' solutions short and elegant. On the other hand, it allows us to explore a number of interesting geometrical problems and pose questions that lead at times to unexpected results and even generalizations.

Let's extend our geometrical interpretation used for proving AM-GM inequality for two positive numbers and examine a few alternative proofs of the following purely algebraic problem:

Prove that for any positive numbers a and b the following inequality is true:

$$\frac{2}{\dfrac{1}{a} + \dfrac{1}{b}} \leq \sqrt{ab} \leq \frac{a + b}{2} \leq \sqrt{\frac{a^2 + b^2}{2}}.$$

PROOF 4.1.

Using a semicircle.

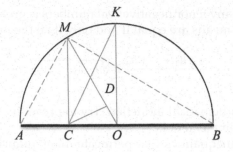

Let's draw a segment AB and mark C on AB such that $AC = a$ and $CB = b$. Next, we will draw a semicircle with center O and diameter AB and construct perpendiculars at C and O till intersection with the semicircle at M and K respectively, $MC \perp AB$ and $KO \perp AB$. Finally, we draw $CD \perp MO$.

This geometrical figure allows us to make the following observations:

1. Being inscribed in a semicircle, $\angle AMB = 90°$. Thus, by the Geometric mean theorem, $MC = \sqrt{ab}$. Also, $MO = OK = OA = OB = \frac{a+b}{2}$ as radii of our semicircle.

2. By construction, triangle MCO is right triangle ($\angle MCO = 90°$), and CD is the altitude dropped to the hypotenuse. Checking similar right triangles ΔMCO and ΔMDC, we see that $\dfrac{MC}{MO} = \dfrac{MD}{MC}$, from which $MC^2 = MD \cdot MO$.

   It implies that $MD = \dfrac{MC^2}{MO} = \dfrac{\left(\sqrt{ab}\right)^2}{\dfrac{a+b}{2}} = \dfrac{2ab}{a+b} = \dfrac{2}{\dfrac{1}{a} + \dfrac{1}{b}}.$

3. $CO = AO - AC = \dfrac{a+b}{2} - a = \dfrac{b-a}{2}.$ Applying the Pythagorean Theorem to triangle COK ($\angle COK = 90°$) gives $CK^2 = CO^2 + OK^2 = \left(\dfrac{b-a}{2}\right)^2 + \left(\dfrac{a+b}{2}\right)^2 =$
   $\dfrac{a^2 + b^2 - 2ab + a^2 + b^2 + 2ab}{4} = \dfrac{a^2 + b^2}{2}.$ Therefore, $CK = \sqrt{\dfrac{a^2 + b^2}{2}}.$

All the preliminary work has been done, and we are ready now to do the comparisons of the four means. Recall that the hypotenuse is the longest side of a right triangle, so thanks to our geometrical interpretation, we can clearly see now, that $MD < MC$ (MD is the leg and MC is the hypotenuse in the right triangle MDC), $MC < MO$ (MC is the leg and MO is the hypotenuse in the right triangle MCO), $MO = OK$ as radii of the same circle and $OK < CK$ (OK is the leg and CK is the hypotenuse in the right triangle KOC); hence $MO < CK$.

All in all, we have $MD < MC < MO < CK$.

It follows that $\dfrac{2}{\dfrac{1}{a} + \dfrac{1}{b}} \le \sqrt{ab} \le \dfrac{a+b}{2} \le \sqrt{\dfrac{a^2 + b^2}{2}}$, as it was required to be proved. When $a = b$, all the segments MD, MC, MO, and KC will coincide with OK. Therefore, all the considered means of a and b will be equal.

PROOF 4.2.

Using circle-secant-tangent relationships.

Relying on a circle's properties, we can get another appealing proof of our inequalities.

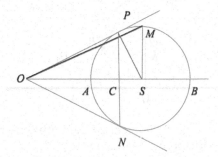

Let's consider three collinear points (they lie on the same straight line) O, A, and B such that $OA = a$, $OB = b$, where $a < b$. Let S be the midpoint of AB. Then $AS = \frac{1}{2} AB = \frac{1}{2}(OB - OA) = \frac{b-a}{2}$, and it follows that $OS = OA + AS = a + \frac{b-a}{2} = \frac{a+b}{2}$. Draw a circle with center at S and diameter AB, and draw at O two straight lines tangent to this circle at points P and N.

By the *Tangent-secant theorem* (you can find the proof in the appendix), $OP = \sqrt{OA \cdot OB} = \sqrt{ab}$. Clearly, considering right triangle OPS ($\angle OPS = 90°$ because $SP \perp OP$ as radius perpendicular to the tangent line at the point of tangency), we see that $OP < OS$, i.e. $\sqrt{ab} < \frac{a+b}{2}$.

Let PN intersect OS at C. Then PC is the altitude in the right triangle OPS, and checking similar triangles OPS and OCP, $\triangle OPS \sim \triangle OCP$, we get that $\frac{OP}{OC} = \frac{OS}{OP}$. It follows that $OC = \frac{OP^2}{OS} = \frac{ab}{\frac{a+b}{2}} = \frac{2ab}{a+b} = \frac{2}{\frac{1}{a} + \frac{1}{b}}$.

Since OC is the leg and OP is the hypotenuse in the right triangle OCP, we see that $OC < OP$, i.e. $\frac{2}{\frac{1}{a} + \frac{1}{b}} < \sqrt{ab}$.

Finally, we draw the radius SM such that $SM \perp OS$. Applying the Pythagorean Theorem to the right triangle OSM gives $OM^2 = OS^2 + SM^2$. Recalling that $OS = \frac{a+b}{2}$ and

$$SM = r = AS = \frac{b-a}{2}, \text{ we have } OM = \sqrt{\left(\frac{a+b}{2}\right)^2 + \left(\frac{b-a}{2}\right)^2} = \sqrt{\frac{2a^2 + 2b^2}{4}} = \sqrt{\frac{a^2 + b^2}{2}}.$$

In the right triangle OSM, $OS < OM$, i.e. $\frac{a+b}{2} < \sqrt{\frac{a^2+b^2}{2}}$. Combining the obtained results of all our calculations, we get the sought-after result

$$\frac{2}{\frac{1}{a} + \frac{1}{b}} \leq \sqrt{ab} \leq \frac{a+b}{2} \leq \sqrt{\frac{a^2+b^2}{2}}.$$

PROOF 4.3.

Using two circles.

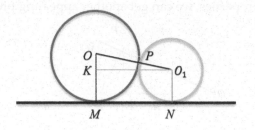

Consider two mutually and externally tangent circles with centers O and O_1 and diameters a and b respectively. Let P be the point of tangency.

Two circles are mutually and externally tangent if the distance between their centers is equal to the sum of their radii, i.e. $OO_1 = OP + PO_1 = \dfrac{a+b}{2}$.

Draw a straight line that is tangent to both circles, and denote the points of tangency M and N. Draw $O_1 K \perp OM$. Observing that $OK = OM - KM = \dfrac{a-b}{2}$, and applying the Pythagorean Theorem to the right triangle OKO_1, gives

$$KO_1 = \sqrt{OO_1^2 - OK^2} = \sqrt{\frac{(a+b)^2}{4} - \frac{(a-b)^2}{4}} = \frac{1}{2}\sqrt{a^2 + 2ab + b^2 - a^2 + 2ab - b^2} = \frac{1}{2}\sqrt{4ab} = \sqrt{ab}.$$

Comparing the hypotenuse OO_1 with the leg KO_1 in the right triangle OKO_1, we see that $OO_1 \geq KO_1$, or in our notations, $\dfrac{a+b}{2} \geq \sqrt{ab}$, and the first part of our proof (the AM-GM inequality) is completed.

It is noteworthy, that by construction, MKO_1N is a rectangle.

Therefore, $MN = KO_1 = \sqrt{ab}$. It means that in passing, we just proved a very useful geometric property:

> *The distance between two points of tangency of two mutually and externally tangent circles with the common tangent straight line is equal to the geometric mean of their diameters.*

Next, we will consider the same configuration of the two mutually and externally tangent circles that have a common tangent straight line as above, but this time, we will drop $PF \perp MN$ and will prove that $PF = \dfrac{ab}{a+b}$.

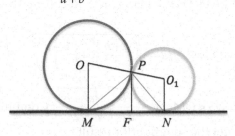

To begin, let's recall that $OP = OM = \dfrac{a}{2}$, $O_1P = O_1N = \dfrac{b}{2}$, and, as we just proved, $MN = \sqrt{ab}$.

Also, for convenience, let's denote equal angles in two isosceles triangles MOP and PO_1N, $\angle OMP = \angle OPM = \alpha$ and $\angle O_1NP = \angle O_1PN = \beta$.

Note that since all three segments OM, PF, and O_1N are perpendicular to MN, they are parallel to each other, $OM \parallel PF \parallel O_1N$. Therefore, in figure above we obtained a right trapezoid MOO_1N, and certainly, confining now ourselves to the properties of this trapezoid makes our arguments simpler and our diagram clearer (see figure below).

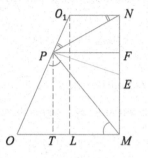

Draw $PT \perp OM$ and $O_1L \perp OM$. From similar right triangles PTO and O_1LO it follows that $\dfrac{OT}{OL} = \dfrac{PO}{O_1O}$. Hence, $OT = \dfrac{OL \cdot PO}{O_1O}$. Notice that $OL = OM - LM = OM - O_1N = \dfrac{a}{2} - \dfrac{b}{2} = \dfrac{a-b}{2}$ and $O_1O = \dfrac{a+b}{2}$. Recalling that $PO = \dfrac{a}{2}$, we can now substitute all values in the last equality and obtain that $OT = \dfrac{\dfrac{a-b}{2} \cdot \dfrac{a}{2}}{\dfrac{a+b}{2}} = \dfrac{a(a-b)}{2(a+b)}$.

By construction, $PFMT$ is a rectangle, hence $PF = TM = OM - OT = \dfrac{a}{2} - \dfrac{a(a-b)}{2(a+b)} = \dfrac{a(a+b) - a(a-b)}{2(a+b)} = \dfrac{a^2 + ab - a^2 + ab}{2(a+b)} = \dfrac{ab}{a+b}$.

Observing the angles formed by point P, we see that $\angle OPM + \angle MPN + \angle O_1PN = 180°$. Thus

$$\angle MPN = 180° - (\alpha + \beta). \tag{4.1}$$

Now, we consider two pairs of equal alternate interior angles inside parallel lines:

$\angle FPM = \angle OMP = \alpha$ (because $OM \parallel PF$),
$\angle FPN = \angle O_1NP = \beta$ (because $PF \parallel O_1N$).

Adding these equalities gives $\angle FPM + \angle FPN = \alpha + \beta$.
 It implies that

$$\angle MPN = \angle FPM + \angle FPN = \alpha + \beta. \tag{4.2}$$

Comparing (4.1) and (4.2) leads to the equation $180° - (\alpha + \beta) = \alpha + \beta$, solving which we get $\alpha + \beta = 90°$. It follows that $\angle MPN = 90°$. So, we obtained that triangle MPN is a right triangle in which PF is its altitude dropped to the hypotenuse MN.

It's not hard to show that $PF \leq \frac{1}{2}MN$. Indeed, connecting P with the mid-point E of MN we form another right triangle PFE ($\angle PFE = 90°$), in which the leg PF is not greater than the hypotenuse PE. But since E is the center of the circumcircle of the right triangle MPN, then $PE = ME = NE = \frac{1}{2}MN$ as radii of the circumcircle, and therefore,

$$PF \leq PE = \frac{1}{2}MN.$$

Recalling now that as it was proved earlier, $MN = \sqrt{ab}$ and $PF = \frac{ab}{a+b}$, after multiplying by 2 numerator and denominator on left side, we arrive at the following inequality:

$$\frac{1}{2} \cdot \frac{2ab}{a+b} \leq \frac{1}{2}\sqrt{ab}.$$

After dividing both sides by $\frac{1}{2}$ it can be rewritten as $\frac{2ab}{a+b} \leq \sqrt{ab}$, which is the inequality relating the harmonic mean and the geometric mean.

Combining our results from the above, we see that $\frac{2ab}{a+b} \leq \sqrt{ab} \leq \frac{a+b}{2}$ or equivalently,

$$\frac{2}{\frac{1}{a}+\frac{1}{b}} \leq \sqrt{ab} \leq \frac{a+b}{2}.$$

It remains to consider the very last step for comparing the arithmetic mean $\frac{a+b}{2}$ with the quadratic mean $\sqrt{\frac{a^2+b^2}{2}}$.

This can be easily done examining two non-intersecting circles that have the common tangent straight line such that the distance between two points of tangency is equal to the arithmetic mean of the diameters of the circles.

Let's consider two circles with centers O_1 and O_2 and diameters a and b respectively, such that after drawing their common tangent MN the distance between M and N is $\frac{a+b}{2}$, $MN = \frac{a+b}{2}$.

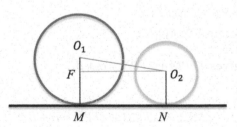

We begin by drawing $O_2F \perp O_1M$. Then in the newly formed rectangle MFO_2N, $FM = O_2N = \frac{b}{2}$ and $O_2F = MN = \frac{a+b}{2}$. It follows that $O_1F = O_1M - FM = \frac{a-b}{2}$.

Consider right triangle $O_2 F O_1$. Applying the Pythagorean Theorem, we have

$$O_2 O_1 = \sqrt{O_1 F^2 + O_2 F^2} = \sqrt{\left(\frac{a-b}{2}\right)^2 + \left(\frac{a+b}{2}\right)^2} = \sqrt{\frac{a^2 - 2ab + b^2 + a^2 + 2ab + b^2}{4}} = \sqrt{\frac{a^2 + b^2}{2}}.$$

In our right triangle $O_2 F O_1$, $O_2 O_1 \geq O_2 F = MN$, which implies that $\sqrt{\dfrac{a^2 + b^2}{2}} \geq \dfrac{a+b}{2}$. We finally arrive at the desired result that

$$\frac{2ab}{a+b} \leq \sqrt{ab} \leq \frac{a+b}{2} \leq \sqrt{\frac{a^2 + b^2}{2}}.$$

PROOF 4.4.

Using a trapezoid and an inscribed circle.

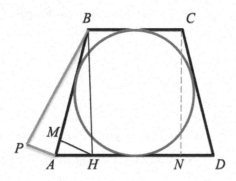

Consider the trapezoid $ABCD$ circumscribed about a circle. Draw $BH \perp AD$ and $HM \perp AB$. Denoting the bases of $ABCD$ as $BC = a$ and $AD = b$, we will prove that $AB = \dfrac{a+b}{2}$, $BH = \sqrt{ab}$, and $BM = \dfrac{2}{\dfrac{1}{a} + \dfrac{1}{b}}$.

First, since our trapezoid is circumscribed about a circle (so-called tangential trapezoid) it is not hard to prove (we leave this to readers) that we are dealing with isosceles trapezoid, i.e., $AB = CD$. The Pitot theorem, named after the French engineer Henri Pitot (1695–1771), states that in a tangential quadrilateral the two sums of lengths of opposite sides are the same. Therefore, $AB + CD = BC + AD$. In our nominations, we have

$$2AB = a + b, \text{ from which } AB = \frac{a+b}{2} \tag{4.3}$$

Second, by drawing $CN \perp AD$, we obtain a right triangle CND congruent to triangle BHA by three sides ($BCNH$ is a rectangle by construction, so $BH = CN$. Also, $AB = CD$ as it was observed earlier, and finally, applying the Pythagorean Theorem to two right triangles, we get the equality of the corresponding remaining sides, $AH = DN$).

Hence, $AH = DN = \dfrac{AD - HN}{2} = \dfrac{b - a}{2}.$

Now, applying the Pythagorean Theorem to the right triangle AHB ($\angle AHB = 90°$),

$$BH^2 = AB^2 - AH^2 = \left(\frac{a+b}{2}\right)^2 - \left(\frac{b-a}{2}\right)^2 = \frac{a^2 + 2ab + b^2 - a^2 + 2ab - b^2}{4} = ab.$$

It implies that, as requested,

$$BH = \sqrt{ab}. \qquad (4.4)$$

Finally, checking similar right triangles BMH and BHA (all respective angles are congruent), we have $\dfrac{BM}{BH} = \dfrac{BH}{BA}$, from which

$$BM = \frac{BH^2}{BA} = \frac{ab}{\dfrac{a+b}{2}} = \frac{2}{\dfrac{1}{a} + \dfrac{1}{b}}. \qquad (4.5)$$

Referring now to right triangles BHM and BHA, we see that $BM \le BH \le AB$. Substituting the respective values from (4.3), (4.4), and (4.5) yields

$$\frac{2}{\dfrac{1}{a} + \dfrac{1}{b}} \le \sqrt{ab} \le \frac{a+b}{2}.$$

We will do now one more auxiliary construction, draw $AP \parallel HM$ such that $AP = AH$.

Connecting P and B, we obtain the right triangle PAB ($\angle PAB = 90°$) in which $AP = AH = \dfrac{b-a}{2}$ and $AB = \dfrac{a+b}{2}$. Applying the Pythagorean Theorem, we get that

$$BP^2 = AP^2 + AB^2 = \left(\frac{b-a}{2}\right)^2 + \left(\frac{a+b}{2}\right)^2 = \frac{2a^2 + 2b^2}{4} = \frac{a^2 + b^2}{2},$$

which leads to $BP = \sqrt{\dfrac{a^2 + b^2}{2}}$. Being the hypotenuse in the right triangle PAB, $BP \ge AB$, i.e. $\sqrt{\dfrac{a^2 + b^2}{2}} \ge \dfrac{a+b}{2}$.

Combining the earlier obtained results with this final inequality, we again arrive at the sought-after conclusion that

$$\frac{2}{\dfrac{1}{a} + \dfrac{1}{b}} \le \sqrt{ab} \le \frac{a+b}{2} \le \sqrt{\frac{a^2 + b^2}{2}}.$$

The next geometrical interpretation of $\dfrac{2}{\dfrac{1}{a} + \dfrac{1}{b}} \le \sqrt{ab} \le \dfrac{a+b}{2} \le \sqrt{\dfrac{a^2 + b^2}{2}}$ allows us not just

to glance at it at a different angle, but to explore several intriguing results not directly related to the considered problem.

Lemma. Given a right triangle ABC ($\angle C = 90°$, $BC = a$, $AC = b$ and $a > b$) and points M and N arbitrarily picked on the hypotenuse AB, prove that N will be located above M if and only if the sum of the distances from N to AC and BC is greater than the sum of the distances from M to AC and BC.

Proof of the Direct Statement

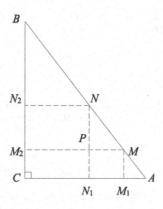

Having N located above M on AB, we need to prove that $NN_1 + NN_2 > MM_1 + MM_2$, where $NN_1 \perp AC$, $NN_2 \perp BC$, $MM_1 \perp AC$, and $MM_2 \perp BC$. Let P be the point of intersection of NN_1 and MM_2. In a right triangle NPM ($\angle P = 90°$), $NP > PM$. Indeed, it is given that $BC = a > AC = b$. Hence, in a right triangle ABC, $\angle A > \angle B$ (a greater side of a triangle is opposite a greater angle – see the proof in the appendix). Obviously, the respective angles in similar triangles ABC and MNP are equal, $\angle M = \angle A$ and $\angle N = \angle B$, so $\angle M > \angle N$, which implies that $NP > PM$. Observe that by construction, $M_2P = NN_2$ and $PN_1 = MM_1$. It follows that

$$NN_1 + NN_2 = (NP + PN_1) + NN_2 = NP + PN_1 + M_2P > PM + MM_1 + M_2P =$$

$$MM_1 + (M_2P + PM) = MM_1 + MM_2.$$

We see that the desired inequality is proved, and indeed, $NN_1 + NN_2 > MM_1 + MM_2$.

Proof of the Converse Statement

It is given to us now that $NN_1 + NN_2 > MM_1 + MM_2$. We have to prove that N lies above M on AB. In fact, our goal is to prove that under the given conditions, in right triangle NPM, angle M is greater than angle N, $\angle M > \angle N$, which is equivalent to the fact that $NP > PM$.

We know that $NN_1 + NN_2 > MM_1 + MM_2$.

Therefore, $NN_1 + NN_2 = NN_2 + NP + PN_1 > M_2P + PM + MM_1$. Cancelling out equal segments' length on each side ($NN_2 = M_2P$ and $PN_1 = MM_1$), we arrive at the sought-after conclusion that $NP > PM$, meaning that indeed, $\angle M > \angle N$. This proves that N lies above M on AB.

Using this Lemma, we will now consider several very specific points on AB and will get the desired relationships.

Case 4.1: *CH* is an angle bisector of angle *C* (*H* ∈ *AB*).

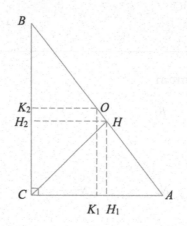

We draw $HH_1 \perp AC$ and $HH_2 \perp BC$. Note that $\triangle BH_2H$ and $\triangle BCA$ are similar right triangles, and thus, their respective sides are in the same proportion, $\dfrac{HH_2}{CA} = \dfrac{BH_2}{BC}$. Recalling the property of an angle bisector that each point on angle bisector is equidistant from the sides of a triangle, we see that CH_2HH_1 is a square. The last equality, therefore, can be rewritten as $\dfrac{HH_2}{CA} = \dfrac{BC - CH_2}{BC} = \dfrac{BC - HH_2}{BC}$. Denoting for convenience $HH_2 = HH_1 = x$, we obtain that $\dfrac{x}{b} = \dfrac{a - x}{a}$. Solving the last equation for x gives $x = \dfrac{ab}{a+b} = \dfrac{1}{\dfrac{1}{a} + \dfrac{1}{b}}$.

By the well-known property, each angle bisector of a triangle divides the opposite side into segments proportional in length to the adjacent sides (see the proof in the appendix). Therefore, $\dfrac{AH}{BH} = \dfrac{AC}{BC} = \dfrac{b}{a}$.

We know that $a > b$, hence, $\dfrac{AH}{BH} < 1$. Also, considering the midpoint O of AB, we have $\dfrac{BO}{AO} = 1$. It implies that O lies above H on AB. Draw perpendiculars OK_1 and OK_2 to AC and BC respectively. Notice that $OK_1 = \dfrac{a}{2}$ and $OK_2 = \dfrac{b}{2}$. Applying now the earlier proved Lemma and comparing the sum of the distances OK_1 and OK_2 from the midpoint O of AB to the legs AC and BC, with the sum of the distances from H to AC and BC, we obtain that $OK_1 + OK_2 > HH_1 + HH_2$, or, in our nominations,

$$\frac{a+b}{2} > \frac{2}{\dfrac{1}{a} + \dfrac{1}{b}}.$$

Case 4.2: Consider the point M on AB such that the sum of the distances from M to the legs of $\triangle ABC$ equals the geometric mean of the legs, $MM_1 + MM_2 = \sqrt{ab}$.

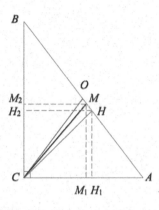

Our goal will be to show that M lies between the midpoint O of AB and point H on AB, the feet of the angle bisector of the right angle C.

We start with drawing $MM_1 \perp AC$ and $MM_2 \perp BC$.

Let's denote $AB = c$ and recall that $BC = a$, $AC = b$ ($a > b$).

Now, observe that in the right triangle ACB we have $\sin A = \dfrac{a}{c}$ and $\sin B = \dfrac{b}{c}$.

Considering the right triangles $MM_1 A$ and $BM_2 M$, and substituting the respective values of $\sin A$ and $\sin B$ from the above, we have

$$MM_1 = AM \cdot \sin A = AM \cdot \frac{a}{c} \text{ and } M_2 M = BM \cdot \sin B = BM \cdot \frac{b}{c}.$$

It is given that $MM_1 + MM_2 = \sqrt{ab}$, therefore, substituting the respective values for MM_1 and MM_2 yields

$$AM \cdot \frac{a}{c} + BM \cdot \frac{b}{c} = \sqrt{ab}.$$

Observing that $BM = c - AM$, and substituting this expression into the last equality gives $AM \cdot \dfrac{a}{c} + (c - AM) \cdot \dfrac{b}{c} = \sqrt{ab}$. Simplifying, we obtain

$$AM \cdot \frac{a}{c} + c \cdot \frac{b}{c} - AM \cdot \frac{b}{c} = \sqrt{ab},$$

$$AM \cdot \frac{a}{c} - AM \cdot \frac{b}{c} + b = \sqrt{ab},$$

$$AM \cdot \left(\frac{a}{c} - \frac{b}{c} \right) = \sqrt{ab} - b,$$

$$AM = \frac{\sqrt{ab} - b}{a - b} \cdot c = \frac{\sqrt{b} \left(\sqrt{a} - \sqrt{b} \right)}{\left(\sqrt{a} - \sqrt{b} \right) \cdot \left(\sqrt{a} + \sqrt{b} \right)} \cdot c = \frac{c \sqrt{b}}{\sqrt{a} + \sqrt{b}} \qquad (4.6)$$

Recall now the results derived in our Case 4.1 that for point H, the feet of the angle bisector of right angle C, $HH_2 = HH_1 = CH_1 = \dfrac{ab}{a+b}$. Therefore, noticing that $H_1A = b - CH_1 = b - HH_2 = b - \dfrac{ab}{a+b} = \dfrac{b^2}{a+b}$ and applying the Pythagorean Theorem to the right triangle HH_1A, we get

$$AH = \sqrt{H_1A^2 + HH_1{}^2} = \sqrt{\left(\frac{b^2}{a+b}\right)^2 + \left(\frac{ab}{a+b}\right)^2} = \sqrt{\frac{b^4 + a^2b^2}{(a+b)^2}} = \sqrt{\frac{b^2\left(b^2 + a^2\right)}{(a+b)^2}}.$$

Noticing that $b^2 + a^2 = c^2$, and that all the numbers a, b, and c are positive, the last equality can be rewritten as

$$AH = \sqrt{\frac{b^2c^2}{(a+b)^2}} = \frac{bc}{a+b} \tag{4.7}$$

To figure out which point, M or H is closer to A, let's find the ratio of AH from (4.7) over AM from (4.6):

$$\frac{AH}{AM} = \frac{\dfrac{bc}{a+b}}{\dfrac{c\sqrt{b}}{\sqrt{a}+\sqrt{b}}} = \frac{bc\left(\sqrt{a}+\sqrt{b}\right)}{c\sqrt{b}(a+b)} = \frac{\sqrt{b}\left(\sqrt{a}+\sqrt{b}\right)}{a+b} = \frac{\sqrt{ab}+b}{a+b}.$$

Recall that for the positive numbers a and b we have $b < a$. Thus $\sqrt{ab} < \sqrt{a^2} = a$.

It follows that in the last ratio, $\dfrac{AH}{AM} = \dfrac{\sqrt{ab}+b}{a+b} < \dfrac{a+b}{a+b} = 1$. It implies that point M lies above H on AB (H is closer to A).

In a similar fashion we now compare the locations of the midpoint O and M on AB. In this case it is easier to find the difference of the distances rather than their ratio:

$$AO - AM = \frac{c}{2} - \frac{c\sqrt{b}}{\sqrt{a}+\sqrt{b}} = \frac{c\left(\sqrt{a}+\sqrt{b}\right) - 2\,c\sqrt{b}}{2\left(\sqrt{a}+\sqrt{b}\right)} = \frac{c\sqrt{a}+c\sqrt{b} - 2\,c\sqrt{b}}{2\left(\sqrt{a}+\sqrt{b}\right)} = \frac{c\sqrt{a}-c\sqrt{b}}{2\left(\sqrt{a}+\sqrt{b}\right)} =$$

$$\frac{c\left(\sqrt{a}-\sqrt{b}\right)}{2\left(\sqrt{a}+\sqrt{b}\right)} > 0.$$

Indeed, the last fraction is positive because its nominator is positive number ($\sqrt{a} - \sqrt{b} > 0$ since $a > b$) and denominator is positive as well (since $\sqrt{a} + \sqrt{b} > 0$).

Therefore, O is above M on AB, and combining the obtained results, we arrive at the conclusion that M lies between O and H on AB. According to the earlier proved Lemma,

$$\frac{2}{\dfrac{1}{a} + \dfrac{1}{b}} \le \sqrt{ab} \le \frac{a+b}{2}.$$

Case 4.3: Consider the point M on AB such that the sum of the distances from M to the legs of $\triangle ABC$ is equal to the quadratic mean of the legs,

$$MM_1 + MM_2 = \sqrt{\frac{a^2+b^2}{2}}.$$

We can use the analysis from Case 4.2 and get that under the conditions for Case 4.3, $AM \cdot \dfrac{a}{c} + BM \cdot \dfrac{b}{c} = \sqrt{\dfrac{a^2+b^2}{2}}$. Doing similar simple manipulations as above and noticing that $a^2 + b^2 = c^2$, we will get to the following equality:

$$AM \cdot \left(\frac{a}{c} - \frac{b}{c}\right) = \sqrt{\frac{a^2+b^2}{2}} - b, \text{ from which } AM = \frac{\frac{c}{\sqrt{2}} - b}{a-b} \cdot c. \ (*)$$

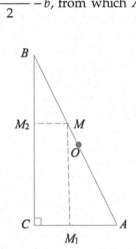

Our goal is to determine which of the two points, M or O (the center of the hypotenuse) lies closer to A on the hypotenuse.

Let's find the difference between $OA = \dfrac{c}{2}$ and AM from (*):

$$OA - AM = \frac{c}{2} - \frac{\frac{c}{\sqrt{2}} - b}{a-b} \cdot c = c \cdot \frac{a - b - c\sqrt{2} + 2b}{2(a-b)} = c \cdot \frac{a + b - c\sqrt{2}}{2(a-b)}.$$

We know that $a > b$, so the denominator of the last fraction is positive. To evaluate the nominator, instead of working with the difference $(a+b) - c\sqrt{2}$ we will find the difference of the squares of two positive numbers $(a+b)$ and $c\sqrt{2}$ (in Chapter 9 we will consider three more alternative proofs of the fact that $(a+b) < c\sqrt{2}$):

$$(a+b)^2 - (c\sqrt{2})^2 = a^2 + 2ab + b^2 - 2c^2 = a^2 + 2ab + b^2 - 2(a^2 + b^2) = a^2 + 2ab + b^2 - 2a^2 - 2b^2 =$$

$$-a^2 + 2ab - b^2 = -(a-b)^2 < 0.$$

When the difference of the squares of two positive numbers is negative, the difference of these two numbers is negative as well. Therefore, our fraction is negative, and respectively,

$OA - AM < 0$. This implies that M lies above O on AB, which in turn leads to $\dfrac{a+b}{2} < \sqrt{\dfrac{a^2+b^2}{2}}$.
Combining the outcomes of all three cases, we arrive at

$$\frac{2}{\dfrac{1}{a}+\dfrac{1}{b}} \le \sqrt{ab} \le \frac{a+b}{2} \le \sqrt{\frac{a^2+b^2}{2}}.$$

In conclusion it should be mentioned that the way we reasoned the locations of points on the hypotenuse involved algebraic inequalities techniques which will be in more detail covered in one of the subsequent chapters. In fact, other than in Case 4.1, we were not concerned at all *how* to locate a point on the hypotenuse, sum of the distances from which to the legs identifies a specific algebraic mean. We only *compared* points' locations in relation to each other, determining which one is above the other one. Another way for approaching this problem may be in figuring out geometrically where each such point has to be on the hypotenuse, i.e., expressing the distance from each point to the end points of the hypotenuse through the legs. Should this be a worthy challenge for the readers to try it on?

I'd like to finish this discussion of the classic algebraic means' geometrical interpretations with one more, perhaps the most popular and frequently mentioned in mathematical texts geometrical interpretation of the above inequalities. All four means can be illustrated as segments in a trapezoid parallel to its bases a and b.

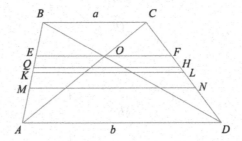

In a trapezoid $ABCD$ with the bases a and b, $EF \parallel QH \parallel KL \parallel MN \parallel BC \parallel AD$.

 EF passes through the point of intersection of the diagonals of the trapezoid $ABCD$.

 QH divides the trapezoid $ABCD$ into two similar trapezoids.

 KL is the middle-line (median) of the trapezoid.

 MN divides the trapezoid $ABCD$ into two trapezoids of equal areas.

Clearly, $EF \le QH \le KL \le MN$. It turns out that these segments represent each respectively the harmonic mean, geometric mean, arithmetic mean, and quadratic mean in terms of the bases a and b of our trapezoid $ABCD$. It's not hard to see that if any two of the segments we are comparing here have the same length, then $ABCD$ becomes a parallelogram and all inequalities convert into equalities.

We invite readers to independently investigate the rigid proofs that the aforementioned segments indeed represent the classic means of the bases of a trapezoid along with the proof of the general case of the considered inequalities for a_i when $a_i > 0$ for any natural number $i > 2$, or do a research in mathematical literature (see the Appendix).

5

Using Algebra for Geometry

Algebra is but written geometry and geometry is but figured algebra.

Marie-Sophie Germain

Elegant and effective solutions can arise when a geometrical problem is translated into the language of algebra. Algebraic methods usually assume introducing variables to "translate" the given conditions into an equation or inequality or system of equations or inequalities. In this chapter, we will examine several traditional, even classical, geometry problems that originally had nothing to do with algebra and demonstrate how important it is to incorporate algebraic methods while searching for their solutions.

We will start with one of the most famous geometrical theorems, the Pythagorean Theorem. There exist more than several hundred proofs, one of which, presented here, illustrates in a lucid and luminous manner amazing links between algebra and geometry.

Originally, the theorem related the areas of the squares constructed on the hypotenuse and legs of a right triangle:

> *The area of the square constructed on the hypotenuse is equal to the sum of the areas of the squares constructed on the legs.*

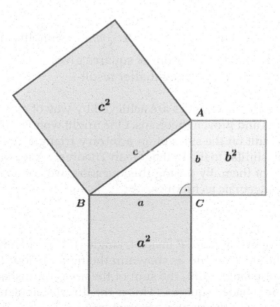

DOI: 10.1201/9781003359500-5

This theorem has an algebraic interpretation through an equation relating the lengths of the sides a, b, and c of a right triangle, and often is called the "Pythagorean equation": $a^2 + b^2 = c^2$, where c represents the length of the hypotenuse and a and b the lengths of the triangle's legs. So, the algebraic formulation of the Pythagorean Theorem is

The square of the hypotenuse is equal to the sum of the squares of the legs.

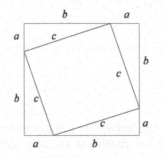

In the figure above, a square that has a side with length $(a + b)$ is divided into four congruent right triangles with sides a, b, and c, and a smaller square with side c. The area of the big square is

$$S = (a+b)^2. \tag{5.1}$$

On the other hand, the same area can be calculated as the sum of the areas of four congruent right triangles and the small square inscribed in the big one. The area of each of our four triangles equals half the product of its legs, $S_\Delta = \dfrac{1}{2}ab$. The area of the inscribed square equals c^2. Therefore, the area of the big square can be derived as

$$S = 4 \cdot \frac{1}{2}ab + c^2 \tag{5.2}$$

Equating (5.1) and (5.2), we obtain that $(a+b)^2 = 4 \cdot \dfrac{1}{2}ab + c^2$. Simplifying the last equality by applying the algebraic formula for the sum of squares gives $a^2 + 2ab + b^2 = 2ab + c^2$, which leads algebraically to $a^2 + b^2 = c^2$, the sought-after result.

Many important results in mathematics are achieved by way of exploring analogies from already solved problems and proven theorems. One might wonder if anything can be concluded about squares built on the sides of an arbitrary triangle, not just a right triangle. Would there be a result similar to the Pythagorean Theorem? A few interesting things can be derived; here is one of them. By posing this question, we got a nice geometrical challenge solvable by pure algebraic techniques.

PROBLEM 5.1.

Three segments are dropped at point P inside triangle ABC perpendicular to its sides. Six squares are built on the sides of ABC as shown in the figure below. Prove that regardless of where P is interior of triangle ABC, the sum of the areas of unshaded squares is equal to the sum of the areas of shaded squares. (This is the direct statement of *Carnot's Theorem*. See Appendix for the proof of the converse Statement).

SOLUTION.

As given, $PM \perp AB$, $PN \perp BC$, and $PK \perp AC$. To simplify the discussion, let's introduce several variables:

$$AM = a,\ MB = b,\ PM = h_1,\ BN = c,\ NC = d,\ PN = h_2,\ CK = m,\ KA = n,\ PK = h_3.$$

Applying the Pythagorean Theorem six times we have the following equations:

From $\triangle AMP$ ($\angle M = 90°$), $AP^2 = a^2 + h_1^2$. From $\triangle AKP$ ($\angle K = 90°$), $AP^2 = n^2 + h_3^2$. Therefore,

$$a^2 + h_1^2 = n^2 + h_3^2 \tag{5.3}$$

From $\triangle BMP$ ($\angle M = 90°$), $BP^2 = b^2 + h_1^2$. From $\triangle BNP$ ($\angle N = 90°$), $BP^2 = c^2 + h_2^2$. Therefore,

$$c^2 + h_2^2 = b^2 + h_1^2 \tag{5.4}$$

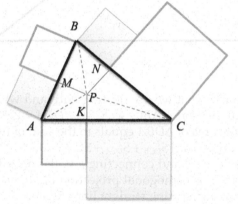

From $\triangle PNC$ ($\angle N = 90°$), $CP^2 = d^2 + h_2^2$. From $\triangle PKC$ ($\angle K = 90°$), $CP^2 = m^2 + h_3^2$. Therefore,

$$m^2 + h_3^2 = d^2 + h_2^2 \tag{5.5}$$

Now, adding (5.3), (5.4), and (5.5) gives

$$a^2 + h_1^2 + c^2 + h_2^2 + m^2 + h_3^2 = n^2 + h_3^2 + b^2 + h_1^2 + d^2 + h_2^2.$$

Cancelling out $h_1^2 + h_2^2 + h_3^2$ on both sides leads to

$$a^2 + c^2 + m^2 = n^2 + b^2 + d^2.$$

Noticing that we have the sum of the areas of the shaded squares on the left-hand side and the sum of the areas of unshaded squares on the right-hand side, we arrive at the desired result.

The Pythagorean Theorem has a three-dimensional analog which is quite often used in applied mathematics:

> *if three faces of a tetrahedron are right triangles with right angles at their common vertex then the sum of the squares of their areas equals the square of the area of the fourth face.*

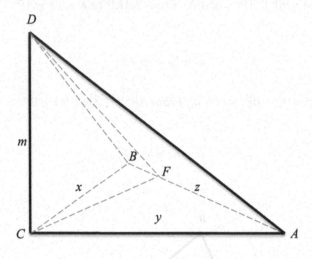

PROOF.

First, let's clearly understand what we are dealing with and what we need to prove. We are given that in the tetrahedron $DABC$, $\angle DCB = \angle DCA = \angle ACB = 90°$. Our goal is to prove that the square of the area of $\triangle DBA$ equals to the sum of the squares of the areas of $\triangle DCB$, $\triangle DCA$, and $\triangle ACB$, i.e., $S_{DBA}^2 = S_{DCB}^2 + S_{DCA}^2 + S_{ACB}^2$.

We start with drawing $CF \perp AB$ and connecting F (F lies on AB) with D. The conditions of the problem imply that CF is orthogonal projection of DF on plane ABC. Indeed, DC, being perpendicular to two intersecting straight lines BC and AC in plane ABC, is perpendicular to this plane and, hence, it is perpendicular to any straight line in that plane. Since BA lies in plane ABC, then $CD \perp AB$. We drew $CF \perp AB$, therefore, by the *Theorem of Three Perpendiculars* (see the appendix), $DF \perp AB$ as well. In other words, as the result of our constructions, DF is the altitude in the triangle DBA.

To simplify calculations we will let $DC = m$, $BC = x$, $AC = y$, $AB = z$, which then allows us to express the areas of our triangles in question through these variables. The area of a triangle is calculated as half the product of its altitude and the base to which it is dropped (half the product of its legs in case of a right triangle). Therefore, the areas of $\triangle DBA$, $\triangle DCB$, $\triangle DCA$, and $\triangle ACB$ can be expressed as following:

$$S_{DBA} = \frac{1}{2}DF \cdot AB \qquad\qquad (5.6)$$

$$S_{DCB} = \frac{1}{2}DC \cdot BC = \frac{1}{2}mx \qquad\qquad (5.7)$$

$$S_{DCA} = \frac{1}{2} DC \cdot AC = \frac{1}{2} my \tag{5.8}$$

$$S_{ACB} = \frac{1}{2} AC \cdot BC = \frac{1}{2} xy. \tag{5.9}$$

Clearly, before we proceed any further, we need to express DF in terms of our variables. First, let's consider $\triangle ACB$ and find its area as $S_{ACB} = \frac{1}{2} CF \cdot AB = \frac{1}{2} CF \cdot z$. On the other hand, we can use its expression from (5.9). Therefore, $\frac{1}{2} CF \cdot z = \frac{1}{2} xy$, which yields $CF = \frac{xy}{z}$.

Now, we will consider the right triangle DCF ($\angle DCF = 90°$) and applying the Pythagorean Theorem find DF as $DF = \sqrt{DC^2 + CF^2} = \sqrt{m^2 + \left(\frac{xy}{z}\right)^2}$. Substituting this expression for DF in (5.6) gives $S_{DBA} = \frac{1}{2} z \cdot \sqrt{m^2 + \left(\frac{xy}{z}\right)^2}$. Squaring both sides leads to

$$S_{DBA}^2 = \frac{1}{4} z^2 \cdot \frac{m^2 z^2 + x^2 y^2}{z^2} = \frac{1}{4}\left(m^2 z^2 + x^2 y^2\right). \tag{5.10}$$

Next, we will square both sides of each equality (5.7), (5.8), and (5.9) and add them. Noticing that in the right triangle ACB, $x^2 + y^2 = z^2$, after adding (5.7), (5.8), and (5.9) we obtain that

$$S_{DCB}^2 + S_{DCA}^2 + S_{ACB}^2 = \frac{1}{4} m^2 x^2 + \frac{1}{4} m^2 y^2 + \frac{1}{4} x^2 y^2 = \frac{1}{4} m^2 \left(x^2 + y^2\right) + \frac{1}{4} x^2 y^2 =$$

$$\frac{1}{4} m^2 z^2 + \frac{1}{4} x^2 y^2 = \frac{1}{4}\left(m^2 z^2 + x^2 y^2\right).$$

Comparing the last equality with (5.10), we arrive at the relationship we set to develop:

$$S_{DBA}^2 = S_{DCB}^2 + S_{DCA}^2 + S_{ACB}^2.$$

Now, we will turn to a geometric construction problem, solution of which leads to interesting generalization, which in turns presents a fruitful technique for solving a pure algebraic problem for solving Diophantine equations (polynomial equations with multiple variables, seeking for integer solutions only).

PROBLEM 5.2.

Is it possible to divide a 19° angle into 19 equal angles using a straightedge and compass?

SOLUTION.

With a compass, construct a circle with center at O, the vertex of our 19° angle, and a radius of any length. Let the points of its intersection with our angle's sides be A and B. Then $\angle AOB = 19°$. Using a compass, we can now draw 18 circles starting with a circle with center at A and radius of AB length. The center of every next circle will be the point of intersection of the previous circle with the originally constructed circle with the center O. In other words, we will divide our original circle into 18 parts. Assume the center of the last one of these circles is C. If we construct one more circle with center at C and the same radius AB, then its intersection with the original circle will be point D such that $\angle DOB = 1°$.

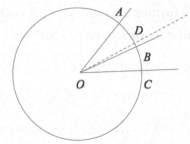

Indeed, since $19 \times 19 = 361$, and we completed the full circle of $360°$ starting and ending at point B, then $\angle DOB = \angle DOC - \angle BOC = 1°$. So, we managed to get $\frac{1}{19}$th of our angle

$\angle AOB = 19°$, and the problem is solved.

We can extend this problem, and consider a general case scenario. Let m and n are two coprime natural numbers and $m < n$. Doing similar constructions as above, and dividing the big circle into $\frac{m}{n}$ equal parts, in n steps we will get m equal arcs. At some x^{th} step we will get to the point on our circle next to the starting point of our journey. We will make several more y full steps to cover $\frac{1}{n}$th of the circle, so $x \cdot \frac{m}{n} = y + \frac{1}{n}$. Solving this equation for x and y, we should be able to answer the question about how a specific angle can be constructed using a straightedge and compass (if our equation has no solutions, then such a construction is impossible). Speaking about mutual benefits, curiously, on the other hand, we just described a *geometrical method* of solving the following algebraic Diophantine equation (it is easily derived from the last equation) in integers:

$$xm - yn = 1.$$

In our original problem, $m = 19$, $n = 360$, $x = 19$, $y = 1$.

In our following explorations we will study several construction problems tackling which algebraic analysis should precede any actual geometric construction attempts.
    We will start with a cute geometrical construction, very simple algebraic analysis of which (perhaps, we can't even dare to call it algebraic, so basic it is) enlightens its solution.

PROBLEM 5.3.

Cut a parallelogram along a straight line through its center so that the two pieces can be rearranged to make a rhombus. (offered by A. Savin in *Quantum*, November/December 1994 issue).

SOLUTION.

Suppose NP is the desired line cutting the given parallelogram $ABCD$ (see Figure 5.1). To get a rhombus, NP has to be equal to BC. Let's see how we can get this line analyzing the

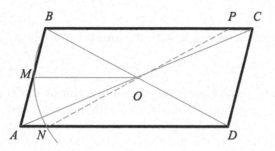

FIGURE 5.1

obtained trapezoid *ABPN*. Let's draw $OM \parallel AD$ (*O* is the point of intersection of the diagonals of *ABCD*). In a trapezoid *ABPN*, *OM* is the midline and its length

$$OM = \frac{1}{2}(BP + AN). \tag{5.11}$$

By Angle-Side-Angle, triangles *AON* and *COP* are congruent ($AO = OC$ as halves of the diagonal *AC*, $\angle AON = \angle COP$ as vertical angles, and $\angle OAN = \angle OCP$ as alternate interior angles inside the parallel lines *AD* and *BC*). Hence, $ON = OP$, and $AN = PC$, which implies that $BC = BP + PC = BP + AN$. Recalling now equality (5.11) we see that in order for *NP* to be equal *BC*, clearly, *ON* has to equal to $OM \left(ON = \frac{1}{2}NP = \frac{1}{2}BC = \frac{1}{2}(BP + AN) = OM \right)$.

To get point *N* on *AD* we need to draw a circle with center *O* and radius *OM*. Its point of intersection with *AD* is the desired point *N*. The last step will be to draw *NO* till its intersection with *BC* at *P*. By construction, $ON = OM = \frac{1}{2}(BP + AN) = \frac{1}{2}BC$, and respectively, $NP = 2ON = BC$. Hence, *NP* is the sought-after line. We can form the new rhombus by shifting *ABPN* over so that *AB* and *DC* coincide (see Figure 5.2).

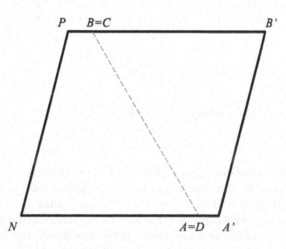

FIGURE 5.2

PROBLEM 5.4.

Inscribe in the given circle a rectangle with the area equal to the area of the given square.

SOLUTION.

Before we start analysis of our problem, let's look a bit at the context of what we are required to do. There is given a circle at center O with radius r. There is also given a square, i.e. we know the length of its side. The goal is to inscribe a rectangle in the given circle. The inscribed rectangle should have the same area as the given square.

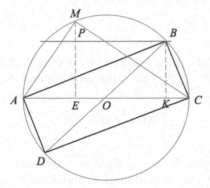

We know that for any inscribed rectangle, its diagonals will be two intersecting diameters of the circle. So, basically, the very first step in our construction is pretty obvious – draw a diameter AC. It has to represent one of the diagonals of our rectangle. Next, locating a point B on our circle in such a way that the area of $\triangle ABC$ equals half the area of the given square, we would get the third vertex, and after drawing straight line BO till its intersection D with our circle we will obtain the fourth vertex of the sought-after rectangle. Since ABC and CDA are two congruent triangles, the area of $ABCD$ will equal to the area of the given square and the conditions of the problem will be satisfied.

So, now the question is how to locate a point B on our circle, such that the area of $\triangle ABC$ equals half the area of the given square. In order to do it, we need to heavily rely on algebraic analysis of the given conditions. Let's introduce several variables: denote $d = 2r$ the diameter of our circle and a the side of the given square. Next, draw $BK \perp AC$, and let $BK = h$. The area of the given square then is expressed as $S_1 = a^2$ and the area of the rectangle will be

$$S_2 = 2S_{ABC} = 2 \cdot \frac{1}{2} AC \cdot BK = d \cdot h.$$

Since by the conditions of the problem, $S_1 = S_2$, then

$$a^2 = dh. \tag{5.12}$$

If we find now point M on the circle, such that $AM = a$ and draw $ME \perp AC$, then AE should have the length of h (to prove this, we will need a few algebra tools from our toolbox!). The next step in our construction will be to find P on ME, such that

$PE = AE = h$, and draw straight line at P parallel to AC intersecting the circle at B.

With this introduction, let's now show, that if we proceed with our constructions, as suggested in the analysis above, then indeed, we would get $AE = h$.

Notice that $\angle AMC = 90°$ because this inscribed angle is subtended by diameter AC. ME is the altitude in the right triangle AMC, therefore, by the Geometric mean theorem, $ME^2 = AE \cdot EC$. To simplify our calculations, we will set $AE = z$, and $EC = y$. Then

$$\begin{cases} z + y = d, \\ z \cdot y = ME^2. \end{cases} \tag{5.13}$$

Let's express ME in terms of a and z, and substitute in Equation (5.13). We will then obtain a simple system with two equations and two variables.

Applying the Pythagorean Theorem to right triangle AEM, we get:
$ME^2 = AM^2 - AE^2 = a^2 - z^2$. Hence, our system is modified to the following

$$\begin{cases} z + y = d, \\ z \cdot y = a^2 - z^2. \end{cases}$$

Substitute $y = d - z$ from the first equation of the system into the second one and perform simple manipulations to get

$$z \cdot (d - z) = a^2 - z^2,$$
$$zd - z^2 = a^2 - z^2,$$
$$zd = a^2.$$

Recalling (5.12), we see that $zd = dh$, therefore $z = AE = h$, which is what we set out to show above.

Segment $AE = h = \dfrac{a^2}{d}$ is constructible, because we know a and d. Therefore, all the next constructions should be easy to perform: locate P on ME, such that $PE = AE = h$, and draw straight line at P parallel to AC intersecting the circle at B.

Concluding our algebraic analysis, we can now pass to the actual construction steps:

1. Draw diameter AC.
2. Draw a circle at center A with radius a (side of the given square) intersecting our given circle at M.
3. Draw $ME \perp AC, E \in AC$.
4. Draw a circle at center E with radius AE intersecting ME at P.
5. Draw a straight line at P parallel to AC intersecting the given circle at B.
6. Draw BO intersecting the given circle at D.
7. Connect points A, B, C, and D. $ABCD$ is the sought-after rectangle.

PROOF.

Since AC and BD are the diameters of the circle, then $ABCD$ is the inscribed rectangle. Its area $S = 2S_{ABC} = AC \cdot BK = dh = a^2$. Hence, $ABCD$ is indeed the desired rectangle.

The problem will have solutions only when $h \leq \dfrac{d}{2}$, which is equivalent to $d \geq a\sqrt{2}$, i.e. the diameter of the given circle has to be greater or equal to the diagonal of the given square.

One final remark – as we see, algebraic analysis of the problem paved the way to our construction decision. Using algebraic technique, the choices of variables should always be chosen judiciously to make the solutions as easy and simple as possible. In this case, introducing variable h was instrumental for performing relatively simple algebraic manipulations leading to the final conclusion. Selecting some other variable perhaps would result in more tedious calculations.

PROBLEM 5.5.

Given that legs of a right triangle are the roots of a quadratic equation
$mx^2 + px + q = 0$ $(m > 0)$, find the inradius of the triangle.

SOLUTION.

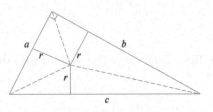

The problem looks tough, so let's try to come up with the plan for the solution, and start with what we know. According to the given conditions, the legs of the right triangle have lengths equal to the roots of the quadratic equation, i.e., $a = x_1$ and $b = x_2$. The goal is to find the radius of the circle inscribed in our triangle. We know that the center of the inscribed circle in a triangle is the point of intersection of triangle's angle bisectors and it is equidistant from all its sides. It implies that each of the perpendiculars dropped from the incenter to each side has the same length as the radius of the inscribed circle.

Connecting the vertices of a triangle to the center of the inscribed circle, we dissect it into three triangles with sides a, b, and c such that their altitudes dropped from the common vertex at the center of the inscribed circle are all equal to the radius r. The area of each of these triangles can be found as half the product of the base by the altitude dropped to that base:

$$S_1 = \frac{1}{2}ar,$$

$$S_2 = \frac{1}{2}br,$$

$$S_3 = \frac{1}{2}cr.$$

Adding the three equalities, we get the area of the original triangle

$$S = S_1 + S_2 + S_3 = \frac{1}{2}ar + \frac{1}{2}br + \frac{1}{2}cr = \frac{1}{2}r(a+b+c) = rp,$$

where $p = \frac{1}{2}(a+b+c)$ is the semi-perimeter of a triangle. We derived a very useful formula, $S = rp$, from which we can express the inradius in question as $r = \frac{S}{p}$. We now see that to calculate r, we need to express the area and semi-perimeter of our triangle in terms of the roots of the given quadratic equation. To do it, referring here to Vieta's formulas (see the appendix) that relate the coefficients of a quadratic equation to sum and product of its roots looks advantageous. We know that the roots of the given quadratic equation satisfy

$$x_1 + x_2 = -\frac{p}{m},$$

$$x_1 \cdot x_2 = \frac{q}{m}.$$

Since we are dealing with the right triangle, then by the Pythagorean Theorem,

$$a^2 + b^2 = c^2.$$

Therefore, squaring the first of Vieta's formulas and substituting the $x_1 \cdot x_2 = \frac{q}{m}$ from the second formula enables us to get $\frac{p^2}{m^2} = (x_1 + x_2)^2 = \underbrace{x_1{}^2 + x_2{}^2}_{c^2} + \underbrace{2\, x_1 \cdot x_2}_{\frac{q}{m}} = c^2 + 2\frac{q}{m}$, from

which $c^2 = \frac{p^2}{m^2} - 2\frac{q}{m}$. Hence, $c = \sqrt{\frac{p^2}{m^2} - 2\frac{q}{m}} = \frac{\sqrt{p^2 - 2qm}}{m}$. The semi-perimeter can be calculated as

$$p = \frac{1}{2}(a+b+c) = \frac{1}{2}(x_1 + x_2 + c) = \frac{1}{2}\left(-\frac{p}{m} + \frac{\sqrt{p^2 - 2qm}}{m}\right) = \frac{-p + \sqrt{p^2 - 2qm}}{2m}. \qquad (5.14)$$

The area of the right triangle equals half the product of its legs, therefore,

$$S = \frac{1}{2} a \cdot b = \frac{1}{2} x_1 \cdot x_2 = \frac{1}{2} \cdot \frac{q}{m} \qquad (5.15)$$

Substituting the expressions from (5.14) and (5.15) into $r = \frac{S}{p}$ gives

$$r = \frac{\dfrac{1}{2} \cdot \dfrac{q}{m}}{\dfrac{-p + \sqrt{p^2 - 2qm}}{2m}} = \frac{q}{\sqrt{p^2 - 2qm} - p},$$

To take one step further and eliminate the irrationality in the denominator, we can multiply both, nominator and denominator, by the conjugate expression to $\left(\sqrt{p^2 - 2qm} - p\right)$ to get

$$r = \frac{q\left(\sqrt{p^2 - 2qm} + p\right)}{\left(\sqrt{p^2 - 2qm} - p\right)\left(\sqrt{p^2 - 2qm} + p\right)} = \frac{q\left(\sqrt{p^2 - 2qm} + p\right)}{p^2 - 2qm - p^2} = \frac{q\left(\sqrt{p^2 - 2qm} + p\right)}{-2qm} = -\frac{\sqrt{p^2 - 2qm} + p}{2m}.$$

So, the sought-after result is $r = -\dfrac{p + \sqrt{p^2 - 2qm}}{2m}$. To finalize the solution, we need to investigate how many solutions this problem has depending on values of m, p, and q.

We leave this to the readers to do such an analysis (it should be also a good exercise working with algebraic inequalities).

What is interesting about this problem is that, in fact, if you are familiar with the property of the inradius that $r = \dfrac{S}{p}$ (it holds for any triangle!), then there is no need even for a diagram to solve this presumably geometrical problem. Vieta's formulas allowed us to get the desired outcome through simple algebraic manipulations with no need to apply any of geometric explorations. Moreover, we gained here a few valuable techniques intertwining algebra and geometry to add to your math toolbox, so you can apply them in similar instances.

PROBLEM 5.6.

This problem was offered in 1980 on Czechoslovakian National math Olympiad.

The greatest side of an isosceles trapezoid is 13 and its perimeter is 28. Find all the sides of the trapezoid if its area is 27. Is it possible that the area of this trapezoid is 27.001?

SOLUTION.

What is interesting about this presumably geometric problem is that not just we are going to implement algebraic techniques for solving it, but even to fully understand the given conditions we will need to do some preliminary algebraic analysis.

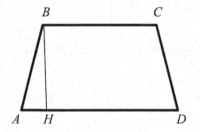

First, let's draw $BH \perp AD$, then BH is the altitude of our trapezoid. Next, assume that AD is the greater of two bases. Then it has to represent the greatest side of the trapezoid and $AD = 13$. Indeed, otherwise, $AB = CD = 13$ (assuming that AD is the greater of two bases but not the greatest side of the trapezoid, we have both of its equal non-parallel sides as the greatest of all sides in the trapezoid), and since we know that perimeter of the trapezoid is 28, then $AD + BC = 28 - 2 \cdot 13 = 2$. The area of a trapezoid is calculated as half the product of the sum of the bases by the altitude. Noticing that in right triangle BHA ($\angle BHA = 90°$), $BH < AB = 13$, we obtain that

$$S_{ABCD} = \frac{1}{2}(AD + BC) \cdot BH < \frac{1}{2} \cdot 2 \cdot 13 = 13 < 27.$$

But this is the contradiction with the given condition that $S_{ABCD} = 27$.

Thus, we are now convinced that the given greatest side is AD and $AD = 13$. Let's denote $AB = CD = x$. Then $BC = 28 - 13 - 2x = 15 - 2x$ and respectively,

$$AH = \frac{1}{2}(AD - BC) = \frac{1}{2}(13 - (15 - 2x)) = \frac{1}{2}(2x - 2) = x - 1.$$

Applying the Pythagorean Theorem to right triangle BHA gives

$$BH = \sqrt{AB^2 - AH^2} = \sqrt{x^2 - (x-1)^2} = \sqrt{x^2 - x^2 + 2x - 1} = \sqrt{2x - 1}.$$

The further analysis entails using the AM-GM inequality for evaluating the area of $ABCD$. Recall that for three positive integers the AM-GM inequality is written as

$$\frac{a_1 + a_2 + a_3}{3} \geq \sqrt[3]{a_1 \cdot a_2 \cdot a_3} \text{ or equivalently, as } \left(\frac{a_1 + a_2 + a_3}{3}\right)^3 \geq a_1 \cdot a_2 \cdot a_3.$$

Setting $14 - x = a_1 = a_2$ and $2x - 1 = a_3$, we have the area of $ABCD$ expressed as the following:

$$S_{ABCD} = \frac{1}{2}(AD + BC) \cdot BH = \frac{1}{2} \cdot (13 + 15 - 2x) \cdot \sqrt{2x - 1} = (14 - x) \cdot \sqrt{2x - 1} =$$

$$\sqrt{(14 - x)^2 \cdot (2x - 1)} \leq \sqrt{\left(\frac{(14 - x) + (14 - x) + (2x - 1)}{3}\right)^3} = \sqrt{\left(\frac{27}{3}\right)^3} = 27.$$

For AM-GM inequality, the equality is possible only when all terms are equal, that is, $14 - x = 2x - 1$, i.e., when $x = 5$. So, we arrive at the lengths of the sides of our trapezoid as $AB = CD = 5$ and $BC = 15 - 2x = 5$. According to our analysis above, it is impossible that the area of this trapezoid is greater than 27, and therefore, it cannot be 27.001, as it was asked in the second question of the problem.

The next problem looks like a pure geometrical construction problem. But we will wonder again at the engaging interweaving of mathematical notions that will arise in our investigations. As in the previous problem, algebraic analysis involving application of AM-GM inequality will be instrumental in getting nice and efficient solution.

PROBLEM 5.7.

Point M lies inside the given angle POQ. Construct the line passing through M such that the sum of the segments it cuts on the sides of the given angle is the smallest one.

SOLUTION.

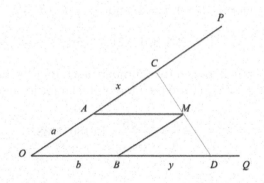

We can draw infinitely many lines through M intersecting the sides of the given angle POQ. To solve the problem, we need to determine how to construct the line CD (let's denote the points of intersection of this line with the sides of the angle C and D) such that $OC + OD$ is the smallest of all such sums.

To analyze the problem, we need two additional constructions: draw $MA \| OQ$ and $MB \| OP$. Also, to facilitate calculations, let's denote $OA = a$, $OB = b$, $AC = x$, and $BD = y$.

Notice that by construction, $OAMB$ is a parallelogram. Thus $MB = OA = a$, and $AM = OB = b$. Checking angles, we see that triangles ACM and BMD are similar, which implies that $\dfrac{AC}{AM} = \dfrac{BM}{BD}$, or in our nominations, $\dfrac{x}{b} = \dfrac{a}{y}$, and equivalently, $xy = ab$.

Now, observe that the sum in question is $OC + OD = OA + AC + OB + BD$. Substituting the respective values, using the just established fact that $xy = ab$, and applying AM-GM inequality to positive numbers x and y, $x + y \geq 2\sqrt{xy}$, gives

$$OC + OD = a + x + b + y = (a + b) + (x + y) \geq a + b + 2\sqrt{xy} = a + b + 2\sqrt{ab} = \left(\sqrt{a} + \sqrt{b}\right)^2.$$

Our algebraic analysis enables us to conclude that the minimal sum has to equal $\left(\sqrt{a} + \sqrt{b}\right)^2$, and this is attained when $x = y = \sqrt{ab}$ (recall that AM-GM inequality $x + y \geq 2\sqrt{xy}$ turns into equality only when $x = y$, and since we got that $xy = ab$, this implies $x = y = \sqrt{ab}$).

Well, it seems we're through.

Having lengths a and b (these are the lengths of the segments on the sides of the given angle after we draw two lines through M parallel to the angle's sides), one can easily construct $AC = BD = \sqrt{ab}$ on each side of the given angle. This can be done referring to strategies explained in Chapter 2. Getting points C and D, the final step will be to draw CD, which according to the analysis above, necessarily will pass through M. CD is the desired segment, which concludes our solution.

There are many more examples to exhibit the power of algebra in facing geometrical topics. As we conclude our discussion, we ought to mention here one of the most remarkable instances of interconnections between these two branches of mathematics: surprising relationships between the famous Fibonacci numbers and the Golden Ratio. The Fibonacci sequence possesses a number of peculiar properties expressed in algebraic terms, whereas in geometry, the Golden Ratio has the reflection in the golden rectangle and golden triangle, pentagon and pentagram; it is broadly used in architecture and art.

To remind, the Fibonacci sequence is a set of natural numbers where the first two terms are 1 and 1 and starting with the third term every number in the sequence equals to the sum of the preceding two terms, $F_1 = 1$, $F_2 = 1$, and $F_{n+1} = F_n + F_{n-1}$:

$$1, 1, 2, 3, 5, 8, 13, 21, 34, 55, \ldots.$$

The Golden Ratio is the ratio between two numbers such that the ratio of the sum of two numbers to the larger of them equals to the ratio of the larger to smaller number.

$$\varphi = \frac{x + y}{y} = \frac{y}{x} = \frac{1 + \sqrt{5}}{2} \approx 1.6180339887$$

$$\overset{\displaystyle x}{\rule{4cm}{0.4pt}}\ \overset{\displaystyle y}{\rule{4cm}{0.4pt}}$$

Amazingly, two seemingly unrelated things are in close relationship to one another, namely, the ratios of Fibonacci successive numbers approximate the golden ratio:

$$\lim_{n\to\infty} \frac{F_{n+1}}{F_n} = \varphi.$$

The Golden ratio is key to understanding properties of the Fibonacci numbers. The detailed investigation of the Golden Ratio and the Fibonacci sequence is beyond the scope of our book. Numerous papers and books are dedicated to the properties of the Golden Ratio and Fibonacci numbers using which the readers can fill in the interest to this subject.

While approaching challenging problems it is crucial to be "mathematically open-minded", i.e., always be prone for utilizing various connections among math disciplines. Speaking about algebra applications in geometry, we can't miss the opportunity to mention a few striking instances of algebraic expressions-formulas in a math discipline dominated by figures and constructions. An algebraically minded reader may find it interesting to trace some formulas presented below to respective geometrical topics covered throughout the book. There are many geometrical problems in which familiarity with algebraically expressed relationships facilitates the solution.

So, in conclusion, here are several well-known formulas (the list can go on and on and on...) for your enjoyment.

Formulas to calculate the area of a triangle (most commonly used):

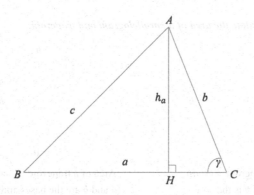

1. $S_{ABC} = \frac{1}{2}ah_a$ – one-half the product of a side by the altitude dropped to that side.

2. $S_{ABC} = \frac{1}{2}ab \cdot \sin\gamma$ – one-half the product of two sides by the sine of an angle between them.

3. *Heron's formula*, sometimes called *Hero's formula*, (after Greek mathematician Heron of Alexandria (c.10–c. 70 AD) which represents the area of a triangle in terms of its sides: $S_{ABC} = \sqrt{p(p-a)(p-b)(p-c)}$, where a, b, and c are the sides and p is the semiperimeter of a triangle, $p = \frac{1}{2}(a+b+c)$.

4. $S_{ABC} = rp$, where r is the radius of the inscribed circle and p is the semiperimeter (see Problem 5.5).

5. $S_{ABC} = \dfrac{abc}{4R}$, where a, b, and c are the sides and R is the radius of the circumscribed circle.

Applying algebraic manipulations with formulas for the area of a triangle, one can get interesting formulas relating triangle's elements. For instance, expressing each altitude from the first formua for the area of a triangle three times we get:

$$h_a = \frac{2S}{a}, \; h_b = \frac{2S}{b}, \; h_c = \frac{2S}{c}. \text{ It implies that } \frac{1}{h_a} = \frac{a}{2S}, \frac{1}{h_b} = \frac{b}{2S}, \text{ and } \frac{1}{h_c} = \frac{c}{2S}.$$

Adding these three equalities and substituting r in terms of S and p from the 4th formula gives:

$$\frac{1}{h_a} + \frac{1}{h_b} + \frac{1}{h_c} = \frac{a+b+c}{2S} = \frac{p}{S} = \frac{1}{r}.$$

So, we arrived at a nice formula relating the altitudes and inradius of a triangle:

$$\frac{1}{h_a} + \frac{1}{h_b} + \frac{1}{h_c} = \frac{1}{r}.$$

Formulas to calculate the area of a parallelogram and trapezoid:

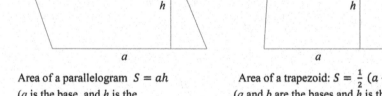

Area of a parallelogram $S = ah$
(a is the base, and h is the altitude dropped to it).

Area of a trapezoid: $S = \frac{1}{2}(a+b)h$
(a and b are the bases and h is the altitude).

General formula for the area of a convex quadrilateral
Bretschneider's formula (after German mathematician Carl Anton Bretschneider (1808–1878):

$$S = \sqrt{(p-a)(p-b)(p-c)(p-d) - abcd \cdot \cos^2\left(\frac{\beta+\gamma}{2}\right)}$$

where a, b, c, d are the sides of the quadrilateral $ABCD$, $\angle ABC = \beta$, $\angle ADC = \gamma$, and p is the semiperimeter, $p = \dfrac{a+b+c+d}{2}$.

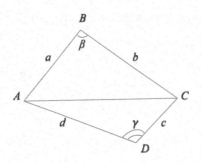

The next three formulas are direct consequences of *Bretschneider's formula*.

Brahmagupta's formula (after the Indian mathematician Brahmagupta (598CE–670CE)) to calculate the area of a cyclic quadrilateral:

$$S = \sqrt{(p-a)(p-b)(p-c)(p-d)}.$$

Area of a quadrilateral that is circumscribed about a circle:

$$S = \sqrt{abcd \cdot \sin^2\left(\frac{\beta+\gamma}{2}\right)}.$$

Area of a quadrilateral that is inscribed in a circle and circumscribed about the circle simultaneously (the area is the square root of the product of its sides):

$$S = \sqrt{abcd}.$$

Algebraic relations in a triangle.

"Extended" *Law of sines*: $\dfrac{a}{\sin\alpha} = \dfrac{b}{\sin\beta} = \dfrac{c}{\sin\gamma} = 2R,$

where a, b, and c, are the sides of a triangle, α, β, and γ are the respective opposite angles, and R is the radius of the triangle's circumcircle.

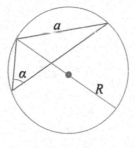

Law of cosines:

$$c^2 = a^2 + b^2 - 2ab \cdot \cos\gamma,$$

where a, b, and c, are the sides of a triangle and γ is the angle between sides a and b.

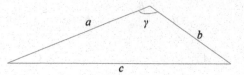

Formula for a median of a triangle in terms of its sides:

$$m_a = \frac{1}{2}\sqrt{2b^2 + 2c^2 - a^2}$$

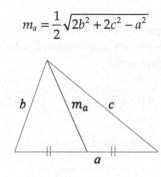

Formula for an angle bisector of a triangle in terms of its sides:

$$l_a = \frac{2bc}{b+c} \cdot \cos\frac{\alpha}{2}$$

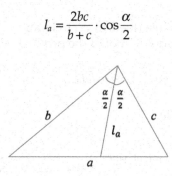

Formula for an angle bisector of a triangle in terms of two sides and two segments on the third side in which it is divided by an angle bisector:

$$l_a = \sqrt{bc - b'c'}$$

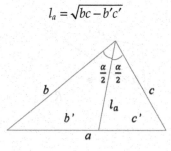

Euler's Formula:

$$d^2 = R^2 - 2Rr$$

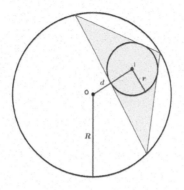

Euler's Formula (after the one of the greatest mathematicians ever lived, Leonhard Euler (1707–1783)) expresses the distance between the centers of the inscribed and circumscribed circles of a triangle in terms of the radii of these circles.

In the formula above, d is the distance between the centers of the inscribed and circumscribed circles of a triangle, r is inradius, and R is circumradius. The proof of this relationship applying inversion properties can be found in the Appendix.

In this chapter we used algebra to explore several geometrical topics. As we evidenced, algebra can shed new light on a deeper appreciation for math relationships. While solving geometric problems, however, in some instances algebraic methods might produce more straightforward and less elegant solutions than applying pure geometric techniques. Depending on a specific problem it is up to you what techniques to apply to get to the desired outcome. The great results usually are achieved when one is well prepared and takes advantage of combinations of multiple methods and techniques for problem solving.

6

Trigonometrical Explorations

Mathematics compares the most diverse phenomena and discovers the secret analogies that unite them.

Joseph Fourier

Trigonometry is encountered in the most diverse problems of algebra, geometry, and calculus. It is a powerful and universal tool in getting links among math disciplines. There are numerous connections between them, expressed by trigonometric identities (you can find such identities in the appendix).

In many cases trigonometry allows us to get appealing and efficient solutions compared to applications of pure geometrical or algebraic techniques.

We will start with trigonometry "servicing" geometrical problems.

PROBLEM 6.1.

Of all the triangles with two sides of the given lengths, find the one with the greatest area.

SOLUTION.

There is no need even to make a diagram to solve this problem, as soon as we applied trigonometry. The area of a triangle can be determined by the formula $S = \frac{1}{2} \cdot a \cdot b \cdot \sin \gamma$, where a and b are the lengths of two sides, and γ is the angle between them. We know that $|\sin \gamma| \leq 1$ for any γ. Hence, the maximum area of a triangle will be attained for $\sin \gamma = 1$, i.e., when $\gamma = 90°$, and it is $S_{max} = \frac{1}{2} \cdot a \cdot b$. So, we conclude that of all the triangles with two sides of the given lengths the right triangle has the greatest area.

PROBLEM 6.2.

In a right triangle, the hypotenuse is c and one of its acute-angle bisectors is $\frac{c\sqrt{3}}{3}$. Find the legs.

SOLUTION.

In the right triangle ABC ($\angle ACB = 90°$), we have $AB = c$ and $AD = \frac{c\sqrt{3}}{3}$ (AD is the angle bisector of $\angle CAB$).

DOI: 10.1201/9781003359500-6

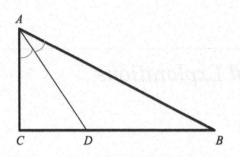

One of the possible approaches to this problem is applying algebraic techniques setting up a system of equations. Let's see where it would get us.

To simplify our calculations, denote $AC = x$, $BC = y$, and $CD = z$. Applying the Pythagorean Theorem to right triangles ACB and ACD, we obtain the following two equations

$$x^2 + y^2 = c^2 \text{ and } x^2 + z^2 = \left(\frac{c\sqrt{3}}{3}\right)^2.$$

Also, we know that an angle bisector of a triangle divides the opposite side into segments proportional in length to the adjacent sides. Therefore, $\dfrac{AC}{AB} = \dfrac{CD}{DB}$, which can be written as $\dfrac{x}{c} = \dfrac{z}{y-z}$.

To summarize, we get the following system of three equations with three variables:

$$\begin{cases} x^2 + y^2 = c^2, \\ x^2 + z^2 = \left(\dfrac{c\sqrt{3}}{3}\right)^2, \\ \dfrac{x}{c} = \dfrac{z}{y-z}. \end{cases}$$

This is not an easy system to solve; it might involve lots of tedious transformations and calculations to get to the result. We leave this exercise to the readers to complete, and will rather concentrate on another much more efficient approach applying trigonometric properties in a right triangle.

Let's denote $\angle CAD = \angle BAD = \alpha$. Then for the right triangle ABC, $AC = c \cdot \cos 2\alpha$.

On the other hand, for the right triangle ACD, $AC = \dfrac{c\sqrt{3}}{3} \cdot \cos\alpha$. Equating the right-hand sides of these two expressions, we get the equation $c \cdot \cos 2\alpha = \dfrac{c\sqrt{3}}{3} \cdot \cos\alpha$.

Let's solve it. Canceling out c and using the formula for cosine of a double angle gives

$$\sqrt{3}\left(\cos^2\alpha - \sin^2\alpha\right) = \cos\alpha.$$

Using the trigonometric *Pythagorean Identity* $\sin^2 \alpha + \cos^2 \alpha = 1$, the last equation easily is simplified to a quadratic equation:

$$\sqrt{3}\left(\cos^2 \alpha - 1 + \cos^2 \alpha\right) = \cos\alpha,$$
$$\sqrt{3}\left(2\cos^2 \alpha - 1\right) = \cos\alpha,$$
$$2\sqrt{3}\cos^2 \alpha - \cos\alpha - \sqrt{3} = 0.$$
$$D = 1 + 4\cdot 2\sqrt{3}\cdot\sqrt{3} = 25. \text{ Therefore, } \cos\alpha = \frac{1\pm\sqrt{25}}{2\cdot 2\sqrt{3}} = \frac{1\pm 5}{4\sqrt{3}}.$$

So, $\cos\alpha = \dfrac{3}{2\sqrt{3}} = \dfrac{\sqrt{3}}{2}$ or $\cos\alpha = -\dfrac{1}{\sqrt{3}}$ (this negative root has to be rejected because α is an acute angle, and only the positive values for $\cos\alpha$ satisfy our problem).

For $\cos\alpha = \dfrac{\sqrt{3}}{2}$, we find $\alpha = 30°$, and respectively, $\angle CAB = 2\alpha = 60°$.

The final steps are to find AC and BC:

$$AC = \frac{c\sqrt{3}}{3}\cdot\cos\alpha = \frac{c\sqrt{3}}{3}\cdot\frac{\sqrt{3}}{2} = \frac{c}{2} \text{ and } BC = c\cdot\sin 2\alpha = c\cdot\sin 60° = \frac{c\sqrt{3}}{2}.$$

As we can see now, trigonometrical approach entails solving just one simple trigonometric equation versus solving a pretty complicated system that we got using an algebraic technique.

PROBLEM 6.3.

Let M and N be the points on sides BC and CD of the square $ABCD$ respectively, such that $BM = \dfrac{1}{2}BC$ and $DN = \dfrac{1}{3}DC$. Find $\angle MAN$.

SOLUTION.

Instead of trying to express $\angle MAN$ through various angles formed inside a square (a pure geometrical approach), we will try to find a trigonometric function of that angle. If we succeed, it will allow us to immediately find the angle in question.

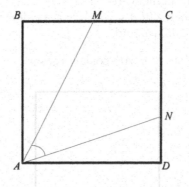

But first some preliminaries.
Notice that $\angle MAN = 90° - (\angle MAB + \angle DAN)$.

Next, consider two right triangles $\triangle ABM$ ($\angle B = 90°$) and $\triangle ADN$ ($\angle D = 90°$). Since we know that $BM = \frac{1}{2}BC$ and $DN = \frac{1}{3}DC$, it makes sense to look for a tangent of angle MAB and tangent of angle DAN as the ratio of the opposite leg to the adjacent leg in each respective right triangle. We get that $\tan \angle MAB = \dfrac{BM}{AB} = \dfrac{1}{2}$ and $\tan \angle DAN = \dfrac{ND}{AD} = \dfrac{1}{3}$.

Now, tangent of our angle can easily be found using the formula for tangent of the sum of two angles.

Recalling that $\tan(x + y) = \dfrac{\tan x + \tan y}{1 - \tan x \tan y}$ and that $\tan(90° - \alpha) = \cot \alpha$, we obtain that

$$\tan \angle MAN = \tan\big(90° - (\angle MAB + \angle DAN)\big) = \cot(\angle MAB + \angle DAN) = \frac{1}{\tan(\angle MAB + \angle DAN)} =$$

$$\frac{1 - \tan \angle MAB \cdot \tan \angle DAN}{\tan \angle MAB + \tan \angle DAN} = \frac{1 - \dfrac{1}{2} \cdot \dfrac{1}{3}}{\dfrac{1}{2} + \dfrac{1}{3}} = \frac{\dfrac{5}{6}}{\dfrac{5}{6}} = 1.$$

Therefore, $\tan \angle MAN = 1$ and we conclude that $\angle MAN = 45°$, which is what we wished to find.

In Problem 6.3 our task was to find an angle measure. This may be considered as a hint to use trigonometry in contemplating a plan for a solution because the request to find an angle measure is equivalent to finding some trigonometric function of that angle.

The application of trigonometry in the following Problem 6.4 is not so obvious, even though the problem somehow resembles the previous one.

PROBLEM 6.4.

Point E lies on side BC of the square $ABCD$, and $AE = a$. AF is the bisector of angle EAD ($F \in CD$). Find $BE + DF$.

SOLUTION.

To solve this problem, we will introduce an auxiliary angle $\varphi = \angle EAD$. For convenience, we also denote the length of the side of the square $AB = x$.

Our goal is to find the sum $BE + DF$. The plan will be to consider a pair of right triangles containing these segments as sides, and try to express each segment in terms of the introduced variables and the given segment $AE = a$.

Notice that since $BC \parallel AD$, then $\angle BEA = \angle DAE = \varphi$ as alternate interior angles.

Considering right triangle ABE, we find

$$BE = AE \cdot \cos\varphi = a\cos\varphi \tag{6.1}$$

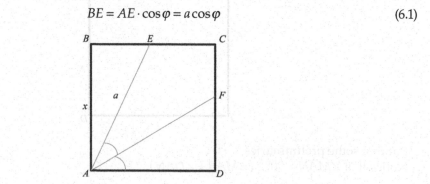

Also,

$$x = AB = a\sin\varphi. \tag{6.2}$$

Next, we consider right triangle ADF. Since $\angle FAD = \dfrac{\varphi}{2}$, we calculate FD as

$$FD = AD \cdot \tan\frac{\varphi}{2} = x\tan\frac{\varphi}{2}.$$

Recalling that $\tan\dfrac{\varphi}{2} = \dfrac{\sin\dfrac{\varphi}{2}}{\cos\dfrac{\varphi}{2}}$ and $\sin\varphi = 2\sin\dfrac{\varphi}{2}\cdot\cos\dfrac{\varphi}{2}$, and substituting $x = AB = AD = a\sin\varphi$

from (6.2) into the last equality gives

$$FD = x\tan\frac{\varphi}{2} = a\sin\varphi \cdot \tan\frac{\varphi}{2} = 2a\sin\frac{\varphi}{2}\cdot\cos\frac{\varphi}{2}\cdot\frac{\sin\dfrac{\varphi}{2}}{\cos\dfrac{\varphi}{2}} = 2a\sin^2\left(\frac{\varphi}{2}\right).$$

Finally, using trigonometric identities $\sin^2\left(\dfrac{\varphi}{2}\right) + \cos^2\left(\dfrac{\varphi}{2}\right) = 1$ and $\cos\varphi = \cos^2\left(\dfrac{\varphi}{2}\right) - \sin^2\left(\dfrac{\varphi}{2}\right)$
and substituting BE from (6.1), gives

$$BE + DF = a\cos\varphi + 2a\sin^2\left(\frac{\varphi}{2}\right) = a\left(\cos^2\left(\frac{\varphi}{2}\right) - \sin^2\left(\frac{\varphi}{2}\right) + 2\sin^2\left(\frac{\varphi}{2}\right)\right)$$

$$= a\left(\cos^2\left(\frac{\varphi}{2}\right) + \sin^2\left(\frac{\varphi}{2}\right)\right) = a\cdot 1 = a.$$

Therefore, we arrive at $BE + DF = a$, which is the answer to the problem.

PROBLEM 6.5.

There is a square inscribed in another square (four vertices of the inscribed square lie on the sides of the circumscribed square). The ratio of their areas is 2:3. Find the value of the smallest angle between the sides of the squares.

SOLUTION.

The square $MNKP$ is inscribed in the square $ABCD$. Knowing that the ratio of their areas is 2:3, $\dfrac{S_{MNKP}}{S_{ABCD}} = \dfrac{2}{3}$, our goal is to find the value of $\angle AMP$.

First, consider the sum of angles by any vertex of $MNKP$. Let's pick, for instance, point N: $\angle BNM + \angle MNK + \angle CNK = 180°$. Therefore, $\angle BNM + 90° + \angle CNK = 180°$, which implies that $\angle BNM + \angle CNK = 90°$. The similar equality holds for angles formed by any other point K, P, and M.

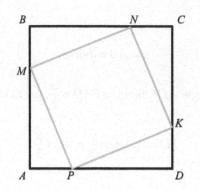

It follows that all the respective angles in four right triangles ΔMBN, ΔNCK, ΔKDP and ΔPAM are equal: $\angle BMN = \angle CNK = \angle DKP = \angle APM$ and $\angle BNM = \angle CKN = \angle KPD = \angle AMP$. Noting that mentioned right triangles have equal hypotenuses (as the sides of the square $MNKP$), we conclude that ΔMBN, ΔNCK, ΔKDP, and ΔPAM are congruent right triangles by Angle-Side-Angle rule. To simplify our calculations, let's denote $AM = BN = CK = DP = x$, $MB = NC = KD = PA = y$, $PM = MN = NK = KP = z$, and $\angle BNM = \alpha$. Our goal is to find α.

Consider right triangle ΔMBN, $BN = x = z\cos\alpha$, $BM = y = z\sin\alpha$. Therefore, the side of the circumscribed square $ABCD$ can be expressed as

$$BC = BN + NC = BN + BM = z\cos\alpha + z\sin\alpha = z(\cos\alpha + \sin\alpha).$$

The Theorem of Ratios of the Areas of Similar Polygons states that the ratio of the areas of the two similar polygons is equal to the squared ratio of their corresponding linear elements. Hence, $\dfrac{S_{MNKP}}{S_{ABCD}} = \dfrac{MN^2}{BC^2} = \dfrac{2}{3}$. Now, substituting the expressions for MN and BC into the last equality gives

$$\frac{z^2}{\left(z(\cos\alpha + \sin\alpha)\right)^2} = \frac{2}{3}.$$

After cancelling out z^2 this is modified to $(\cos\alpha + \sin\alpha)^2 = \dfrac{3}{2}$.

Squaring and making simplifications using the Pythagorean Identity $\sin^2\alpha + \cos^2\alpha = 1$ and the formula $2\cos\alpha \cdot \sin\alpha = \sin 2\alpha$ gives

$$\cos^2\alpha + \sin^2\alpha + 2\cos\alpha \cdot \sin\alpha = \frac{3}{2},$$

$$1 + \sin 2\alpha = \frac{3}{2},$$

$$\sin 2\alpha = \frac{1}{2}.$$

Clearly, $0° < \alpha < 45°$, therefore from $\sin 2\alpha = \dfrac{1}{2}$ it follows that $2\alpha = 30°$ and respectively, $\alpha = 15°$, which is the result we seek.

PROBLEM 6.6.

Knowing that two altitudes in an acute-angled triangle ABC are not shorter than the sides to which they are dropped, $h_c \geq AB$ and $h_a \geq BC$, find the angles of $\triangle ABC$.

<u>SOLUTION.</u>

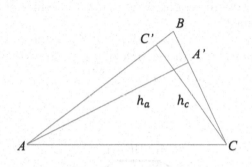

First, we have to note that our diagram is not drawn to scale, as we have no idea what the angles measures are (the goal is to find them!).

We know that $AA' \perp BC$, $CC' \perp AB$ and $AA' = h_a \geq BC$, $CC' = h_c \geq AB$. Not much is given to us at all. It is not clear how this info may lead to determine the angles measures. Logically, we may try analyzing the obtained two right triangles $AA'B$ and $CC'B$ and see if we can come up with some useful ideas.

In the right triangle $AA'B$,

$$\sin B = \frac{AA'}{AB} = \frac{h_a}{AB} \geq \frac{h_a}{h_c} \qquad (6.3)$$

In the right triangle $CC'B$,

$$\sin B = \frac{CC'}{BC} = \frac{h_c}{BC} \geq \frac{h_c}{h_a} \qquad (6.4)$$

Dealing with a triangle, we know that any of its angles is less then $180°$, so $\sin B > 0$.

Clearly, $h_a > 0$ and $h_c > 0$ as well. Therefore, multiplying (6.3) and (6.4) yields

$$\sin^2 B \geq \frac{h_a}{h_c} \cdot \frac{h_c}{h_a} = 1.$$

From the last inequality we conclude that $\sin B = 1$ (by the properties of the sine function, $|\sin x| \leq 1$ for any angle x). It implies that under the given conditions, $\angle B = 90°$. Furthermore, it follows that AA' coincides with AB and CC' coincides with BC, i.e. $AB = h_a$ and $BC = h_c$ ($AB \perp BC$). Recall now that as given, $h_a \geq BC$ and $h_c \geq AB$. So, $AB \geq BC$ and $BC \geq AB$ at the same time. We arrive at the only possible outcome that $AB = BC$. Hence, triangle ABC turns out to be a right isosceles triangle with $\angle B = 90°$ and $\angle A = \angle C = 45°$.

Referring to the properties of the sine function allowed us to come up with nice and elegant solution of seemingly (at least at the first glance) difficult problem.

PROBLEM 6.7.

A regular 12-gon $A_1 A_2 \ldots A_{12}$ is circumscribed about the circle with radius r. Prove that $A_1 A_2 + A_1 A_4 = A_1 A_6 = 2r$.

SOLUTION.

Let's denote O the center of our regular 12-gon; it is the center of the circumscribed and inscribed circle at the same time.

In our diagram below, $OM = r$ (where $OM \perp A_1 A_2$) is the radius of the incircle and $OA_1 = OA_2 = R$ is the radius of the circumcircle of the given regular 12-gon.

The value of the central angle $\angle A_1 O A_2 = \dfrac{360°}{12} = 30°$, $\angle A_1 O A_4 = \dfrac{360°}{12} \cdot 3 = 90°$, and $\angle A_1 O A_6 = \dfrac{360°}{12} \cdot 5 = 150°$.

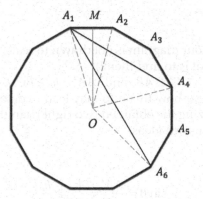

It is easy to prove that for a chord of a circumcircle of a regular polygon with radius R, the length of a chord is expressed in terms of R and the trigonometric function of the central angle α subtended by this chord as $x = 2R \cdot \sin \dfrac{\alpha}{2}$ (it is derived from a right triangle obtained after drawing an altitude to the base of isosceles triangle formed connecting the center of a circle with the end points of a chord).

It follows that $A_1 A_2 = 2R \cdot \sin 15°$, $A_1 A_4 = 2R \cdot \sin 45°$, and $A_1 A_6 = 2R \cdot \sin 75°$. Remember, our goal is to prove that $A_1 A_2 + A_1 A_4 = A_1 A_6$. Substituting the values from above, our goal is modified to proving that $2R \cdot \sin 15° + 2R \cdot \sin 45° = 2R \cdot \sin 75°$, and after canceling out $2R$ on both sides, our problem is now reduced to proving the following trigonometric equality:

$$\sin 15° + \sin 45° = \sin 75°.$$

We can modify the left side using the formula $\sin x + \sin y = 2 \sin \dfrac{x+y}{2} \cdot \cos \dfrac{x-y}{2}$ and get

$$\sin 15° + \sin 45° = 2 \sin \frac{15° + 45°}{2} \cdot \cos \frac{15° - 45°}{2} = 2 \sin 30° \cdot \cos(-15°) = 2 \cdot \frac{1}{2} \cdot \cos 15° = \cos 15°.$$

On the other hand, the expression on the right side is easily modified as

$$\sin 75° = \sin(90° - 15°) = \cos 15°.$$

We see that $\cos 15° = \cos 15°$, which proves that indeed, $A_1A_2 + A_1A_4 = A_1A_6$, as it was required to be proved, and we are done with the first part of our problem.

The next step will be to prove that $A_1A_6 = 2r$.

In the right triangle OMA_1 we have $OA_1 = R$ (radius of the circumcircle of our 12-gon), $OM = r$ (radius of the incircle of our 12-gon), and $\angle OA_1M = 75°$ (in isosceles $\triangle OA_1A_2$, $\angle A_1 = \angle A_2 = \dfrac{180° - 30°}{2} = 75°$). Hence, $R = \dfrac{r}{\sin 75°}$.

We can calculate $\sin 75°$ using trigonometric formula for the sine of the sum of two angles, $\sin(x+y) = \sin x \cdot \cos y + \sin y \cdot \cos x$, and get that

$$\sin 75° = \sin(45° + 30°) = \sin 45° \cdot \cos 30° + \sin 30° \cdot \cos 45° = \frac{\sqrt{2}}{2} \cdot \frac{\sqrt{3}}{2} + \frac{1}{2} \cdot \frac{\sqrt{2}}{2} = \frac{\sqrt{2}}{4}\left(\sqrt{3} + 1\right).$$

So, $R = \dfrac{r}{\sin 75°} = \dfrac{4r}{\sqrt{2}\left(\sqrt{3}+1\right)}$ (6.5)

Now, we can apply the *Law of cosines* to $\triangle A_1OA_6$ and find A_1A_6:

$$A_1A_6{}^2 = A_1O^2 + A_6O^2 - 2A_1O \cdot A_6O \cdot \cos \angle A_1OA_6.$$

Note that $\angle A_1OA_6 = 150°$.

Therefore,

$$A_1A_6{}^2 = R^2 + R^2 - 2R^2 \cdot \cos 150° = 2R^2\left(1 - \cos(180° - 30°)\right) =$$

$$2R^2\left(1 + \cos 30°\right) = 2R^2\left(1 + \frac{\sqrt{3}}{2}\right) = R^2\left(2 + \sqrt{3}\right).$$

Substituting R from (6.5) into the last expression gives

$$A_1A_6{}^2 = \left(\frac{4r}{\sqrt{2}\left(\sqrt{3}+1\right)}\right)^2 \cdot \left(2 + \sqrt{3}\right) = \frac{8r^2}{\left(\sqrt{3}+1\right)^2} \cdot \left(2 + \sqrt{3}\right) = \frac{8r^2}{3 + 2\sqrt{3} + 1} \cdot \left(2 + \sqrt{3}\right) =$$

$$\frac{8r^2}{2\left(2+\sqrt{3}\right)} \cdot \left(2 + \sqrt{3}\right) = 4r^2,$$

from which we arrive at $A_1A_6 = 2r$, as it was required to be proved.

In the last problem instead of seeking for a proof by applying conventional pure geometrical techniques, we transformed the problem to proving a trigonometric identity. After applying the Law of cosines and doing just a few trigonometric simplifications, we also managed to prove that A_1A_6 equals to the diameter of the incircle of the given regular 12-gon.

PROBLEM 6.8.

There is given a rectangle $ABCD$ with sides $AB = a$ and $AD = b$. It is circumscribed by rectangles in such a way, that each side of every circumscribed rectangle passes through one of the vertices of $ABCD$. Determine which one of these rectangles has the greatest area and find its area.

SOLUTION.

Denote by φ the angle between the sides of the given rectangle $ABCD$ and arbitrary circumscribed rectangle $MNKL$, $\angle ABM = \varphi$. Since $\angle ABC = 90°$, then

$$\angle NBC = 180° - 90° - \varphi = 90° - \varphi.$$

In the right triangle AMB ($\angle M = 90°$), $MB = a \cdot \cos\varphi$ and $MA = a \cdot \sin\varphi$.

In the right triangle BNC ($\angle N = 90°$), $BN = b \cdot \cos(90° - \varphi) = b \cdot \sin\varphi$ and $NC = b \cdot \sin(90° - \varphi) = b \cdot \cos\varphi$.

From the congruency of the right triangles BNC and DLA (they have equal hypotenuses and two respective equal angles by each hypotenuse) it follows that $BN = DL$ and $NC = AL$. Therefore, the sides of $MNKL$ can be expressed as the following:

$$MN = MB + BN = a \cdot \cos\varphi + b \cdot \sin\varphi \text{ and } ML = MA + AL = a \cdot \sin\varphi + b \cdot \cos\varphi.$$

The area of the rectangle $MNKL$ is calculated as the product of its sides:

$$S_{MNKL} = MN \cdot ML = (a \cdot \cos\varphi + b \cdot \sin\varphi) \cdot (a \cdot \sin\varphi + b \cdot \cos\varphi) =$$
$$a^2 \cos\varphi \cdot \sin\varphi + ab\cos^2\varphi + ab\sin^2\varphi + b^2 \cos\varphi \cdot \sin\varphi =$$
$$(a^2 + b^2)\cos\varphi \cdot \sin\varphi + ab(\cos^2\varphi + \sin^2\varphi) =$$
$$(a^2 + b^2)\cos\varphi \cdot \sin\varphi + ab.$$

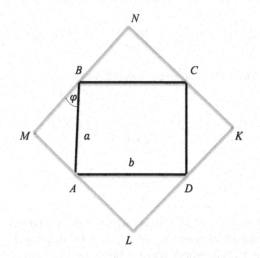

In the last expression we have to determine the maximum value dependent on angle φ. This can be easily done by applying AM-GM inequality to the product $\cos\varphi \cdot \sin\varphi$:

$$\cos\varphi \cdot \sin\varphi = \sqrt{\cos^2\varphi \cdot \sin^2\varphi} \le \frac{1}{2}(\cos^2\varphi + \sin^2\varphi) = \frac{1}{2} \cdot 1 = \frac{1}{2}.$$

The equality will be attained when $\cos\varphi = \sin\varphi = \dfrac{1}{\sqrt{2}}$, i.e. when $\varphi = 45°$ (because φ was selected such that $0° < \varphi < 90°$). It implies that out of all circumscribed rectangles, the one with the greatest area will be the rectangle with sides forming the angles of 45° with the sides of the given rectangle *ABCD*. In this case all four right triangles *AMB*, *BNC*, *CKD*, and *DLA* become congruent isosceles triangles, meaning that such a rectangle turns out to be a square. Its area is calculated as

$$S_{max} = \left(a^2 + b^2\right)\cos 45° \cdot \sin 45° + ab = \left(a^2 + b^2\right)\frac{\sqrt{2}}{2} \cdot \frac{\sqrt{2}}{2} + ab = \frac{1}{2}\left(a^2 + b^2 + 2ab\right) = \frac{1}{2}\left(a+b\right)^2,$$

which concludes our solution.

As we exhibited the usefulness of trigonometric manipulations in tackling geometrical problems, now it's time to reveal that sometimes, geometry returns a favor and can be efficiently utilized to solve pure trigonometric problems.

PROBLEM 6.9.

Prove that $\sin(\alpha + \beta) = \sin \alpha \cos \beta + \cos \alpha \sin \beta$.

SOLUTION.

Consider two right triangles *ACB* and *DCB* (the right angle is at vertex *C* in each triangle) with the common leg *BC* and angles $\angle ABC = \alpha$ and $\angle DBC = \beta$.

These two triangles form another triangle *ABD* with the angle $\angle ABD = \alpha + \beta$.

The area of $\triangle ABD$ can be calculated as $S_{ABD} = \frac{1}{2}AB \cdot BD \cdot \sin(\alpha + \beta)$.

The area of $\triangle ABC$ can be calculated as $S_{ABC} = \frac{1}{2}AB \cdot BC \cdot \sin \alpha$;

The area of $\triangle DBC$ can be calculated as $S_{DBC} = \frac{1}{2}BD \cdot BC \cdot \sin \beta$.

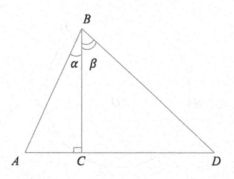

Since the area of $\triangle ABD$ equals the sum of the areas of $\triangle ABC$ and $\triangle DBC$, we see that $\frac{1}{2}AB \cdot BD \cdot \sin(\alpha + \beta) = \frac{1}{2}AB \cdot BC \cdot \sin \alpha + \frac{1}{2}BD \cdot BC \cdot \sin \beta$, and after dividing both sides by $\frac{1}{2}AB \cdot BD$, we obtain that

$$\sin(\alpha + \beta) = \frac{BC}{BD} \cdot \sin \alpha + \frac{BC}{AB} \cdot \sin \beta. \tag{6.6}$$

In the right triangle ACB, $\dfrac{BC}{AB} = \cos\alpha$ and in the right triangle BCD, $\dfrac{BC}{BD} = \cos\beta$. Substituting these expressions into (6.6) gives us the desired result:

$$\sin(\alpha + \beta) = \sin\alpha\cos\beta + \cos\alpha\sin\beta.$$

Another nice alternative to derive the above formula is to consider a right triangle with the hypotenuse 1 and angle β inscribed in a square as depicted in a figure below.

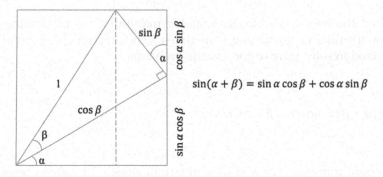

In Chapter 3 we got acquainted with Ptolemy's theorem. This theorem can be conveniently used to derive another trigonometric identity.

PROBLEM 6.10.

Prove that $\sin(\alpha - \beta) = \sin\alpha\cos\beta - \cos\alpha\sin\beta$.

SOLUTION.

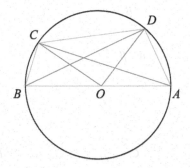

We preface the proof of our identity with several observations.

We will consider a quadrilateral $ABCD$ inscribed into a circle with center O, such that its side AB is the diameter of the circle.

Let $\angle AOC = 2\alpha$ and $\angle AOD = 2\beta$. By the Inscribed angle theorem,

$$\angle ABC = \frac{1}{2}\cdot\angle AOC = \frac{1}{2}\cdot 2\alpha = \alpha \text{ and } \angle ABD = \frac{1}{2}\cdot\angle AOD = \frac{1}{2}\cdot 2\beta = \beta.$$

Applying the Law of sines to triangles ABC and ABD and denoting R the radius of our circle, gives $\dfrac{AC}{\sin\alpha} = 2R$ and $\dfrac{AD}{\sin\beta} = 2R$. Thus, $AC = 2R\cdot\sin\alpha$ and $AD = 2R\cdot\sin\beta$.

Next, we notice that $\angle BOC = 180° - \angle AOC = 180° - 2\alpha$.

Respectively, by the Inscribed angle theorem, $\angle BAC = \dfrac{1}{2}\cdot\angle BOC = \dfrac{1}{2}\cdot(180° - 2\alpha) = 90° - \alpha$.

Therefore, using the Law of sines to triangle ABC one more time, $\dfrac{BC}{\sin(90° - \alpha)} = 2R$, from which $BC = 2R\cdot\sin(90° - \alpha) = 2R\cdot\cos\alpha$.

Similarly, $\angle BOD = 180° - \angle AOD = 180° - 2\beta$.

Respectively, $\angle BAD = \dfrac{1}{2}\cdot\angle BOD = \dfrac{1}{2}\cdot(180° - 2\beta) = 90° - \beta$.

Therefore, $\dfrac{BD}{\sin(90° - \beta)} = 2R$, from which $BD = 2R\cdot\sin(90° - \beta) = 2R\cdot\cos\beta$.

Finally, notice that $\angle COD = \angle AOC - \angle AOD = 2\alpha - 2\beta$.

Hence, $\angle CAD = \dfrac{1}{2}\cdot\angle COD = \alpha - \beta$. Applying the Law of sines to triangle CAD, yields $\dfrac{CD}{\sin(\alpha - \beta)} = 2R$, from which $CD = 2R\cdot\sin(\alpha - \beta)$.

All our preliminary work has been done, and now the desired formula is rendered by a straightforward application of Ptolemy's theorem which asserts that the following equality holds true:

$$AC\cdot BD = AB\cdot CD + AD\cdot BC.$$

Substituting the values of all segments from the above calculations, and recalling that $AB = 2R$ gives

$$(2R\cdot\sin\alpha)\cdot(2R\cdot\cos\beta) = 2R\cdot 2R\cdot\sin(\alpha - \beta) + (2R\cdot\sin\beta)\cdot(2R\cdot\cos\alpha)$$

Dividing both sides by $4R^2$ and simplifying, we obtain the sought-after formula

$$\sin(\alpha - \beta) = \sin\alpha\cos\beta - \cos\alpha\sin\beta.$$

PROBLEM 6.11.

Prove that $\cos 36° - \cos 72° = \dfrac{1}{2}$.

SOLUTION.

We suggest that the readers investigate the trigonometric proof of this equality independently. For instance, you can multiply both sides by $\sin 36°$ and do the modest simplifications to get the desired result. As an alternative, we will examine the following simple and elegant geometrical solution of this problem.

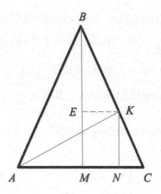

We will consider an isosceles triangle ABC ($AB = CB$) such that $\angle ABC = 36°$ and $AC = 1$. Notice that since $\angle ABC = 36°$ then $\angle BAC = \angle BCA = 72°$.

Draw the angle bisector AK. Then $\angle BAK = \angle CAK = 36°$ and, therefore, in isosceles triangle ABK, $AK = BK$. Also, in triangle KAC we have $\angle AKC = 180° - 36° - 72° = 72°$. Hence, this triangle is isosceles as well, and $AK = AC = 1$. Referring to equality $AK = BK$, we see that $BK = AK = AC = 1$.

Now, we will drop $BM \perp AC$ and $KN \perp AC$.

Notice that in the right triangle ANK,

$$AN = AK \cdot \cos 36° = \cos 36°. \tag{6.7}$$

Next, draw $KE \perp BM$, ($E \in BM$).

Then in the right triangle BEK, $EK = BK \cdot \cos 72° = 1 \cdot \cos 72° = \cos 72°$. By construction, $EKNM$ is a rectangle, therefore,

$$MN = EK = \cos 72°. \tag{6.8}$$

To conclude our proof, we merely have to observe that $AM = \frac{1}{2}AC = \frac{1}{2}$ (because BM being the altitude in the isosceles triangle dropped to its base, is its median at the same time) and that $AM = AN - MN$. Substituting the values from (6.7) and (6.8) gives us the sought-after equality $\cos 36° - \cos 72° = \frac{1}{2}$.

PROBLEM 6.12.

Prove that if $\angle A + \angle B + \angle C = 180°$ then the following equality holds:

$$\cot\frac{A}{2} + \cot\frac{B}{2} + \cot\frac{C}{2} = \cot\frac{A}{2} \cdot \cot\frac{B}{2} \cdot \cot\frac{C}{2}.$$

SOLUTION.

A trigonometric proof of this equality is not easy to find. We leave it to explore to the readers. Instead, noticing that the sum of angles in a triangle is equal to 180°, we consider here a different approach to this problem and employ a geometrical interpretation. It would allow us to glance at several notable relationships in a triangle that are worth adding to your math toolbox. Also, by using this strategy, we would overcome the problem without boring and painstaking trigonometrical calculations.

Consider a random $\triangle ABC$, denote O its incenter (the center of the inscribed circle in a triangle, the point where the internal angle bisectors of the triangle cross; it is the point equidistant from the triangle's sides) and drop the perpendiculars OK, OM, and ON to AB, BC, and AC respectively.

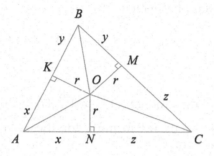

Connecting O with three vertices A, B, and C we form three pairs of congruent right triangles, $\triangle AKO \cong \triangle ANO$, $\triangle BKO \cong \triangle BMO$, and $\triangle CMO \cong \triangle CNO$ (by the common hypotenuse and leg representing the inradius of the triangle).

To simplify our calculations, let's denote by r the distance from O to each side (it is the inradius – radius of the inscribed circle), and by x, y, and z the segments in which each side is divided by the feet of the respective radius:

$$AK = AN = x, \ BK = BM = y, \text{ and } CM = CN = z.$$

In order to proceed with the proof of our equality, we first derive a few necessary formulas.

If we denote the sides of our triangle $BC = a$, $AC = b$, $AB = c$, and its semi-perimeter as p, then $p = \dfrac{a+b+c}{2}$. Clearly, in our nominations, since

$$\begin{cases} x+y=c, \\ y+z=a, \\ x+z=b, \end{cases}$$

we see that $x + y + y + z + x + z = c + a + b$, from which

$$x+y+z = \frac{a+b+c}{2} = p. \tag{6.9}$$

By adding the first and the third equations in the system above, we get $2x+(y+z)=c+b$, and substituting the value of a from the second equation, we obtain that $x = \dfrac{b+c-a}{2} = \dfrac{b+c+a-2a}{2} = p - \dfrac{2a}{2} = p-a.$

In a similar way, $y = p - b$ and $z = p - c$.

Next, we will consider the right triangle AKO ($\angle K = 90°$). Recalling that AO is the angle bisector of $\angle A$ (O, being the triangle's incenter, is the point of intersection of its internal angle bisectors), we have $\angle KAO = \frac{1}{2}\angle BAC = \frac{A}{2}$. So, for $\triangle AKO$, it follows that

$\cot\frac{A}{2} = \frac{AK}{KO} = \frac{p-a}{r}$. In the same way we can get that $\cot\frac{B}{2} = \frac{p-b}{r}$ and $\cot\frac{C}{2} = \frac{p-c}{r}$.

Let's now find the sum $\left(\cot\frac{A}{2} + \cot\frac{B}{2} + \cot\frac{C}{2}\right)$ substituting for each addend the expression through the sides and semi-perimeter:

$$\cot\frac{A}{2} + \cot\frac{B}{2} + \cot\frac{C}{2} = \frac{p-a}{r} + \frac{p-b}{r} + \frac{p-c}{r} = \frac{3p-(a+b+c)}{r} = \frac{3p-2p}{r} = \frac{p}{r} \qquad (6.10)$$

Now, our goal will be to find the product $\cot\frac{A}{2} \cdot \cot\frac{B}{2} \cdot \cot\frac{C}{2}$ and compare it to (6.10).

Here, we need to recall the useful formula for the area of a triangle derived in Problem 5.5 of the previous chapter:

$$S = rp. \qquad (6.11)$$

Also, by Heron's formula, the area of $\triangle ABC$ can be found as

$$S = \sqrt{p(p-a)(p-b)(p-c)}, \text{ from which } \frac{S^2}{p} = (p-a)(p-b)(p-c) \qquad (6.12)$$

We are now done with all the preliminary observations, and by substituting the values from (6.11) and (6.12), we proceed to find the expression for the product $\cot\frac{A}{2} \cdot \cot\frac{B}{2} \cdot \cot\frac{C}{2}$:

$$\cot\frac{A}{2} \cdot \cot\frac{B}{2} \cdot \cot\frac{C}{2} = \frac{p-a}{r} \cdot \frac{p-b}{r} \cdot \frac{p-c}{r} = \frac{(p-a)(p-b)(p-c)}{r^3} = \frac{S^2}{p} \cdot \frac{1}{r^3} = \frac{(rp)^2}{p} \cdot \frac{1}{r^3} = \frac{p}{r}.$$

Comparing the last equality with the equality (6.10), we arrive at the desired conclusion that

$$\cot\frac{A}{2} + \cot\frac{B}{2} + \cot\frac{C}{2} = \cot\frac{A}{2} \cdot \cot\frac{B}{2} \cdot \cot\frac{C}{2}.$$

Trigonometry applications are also useful in simplifying solutions to many algebraic problems. Sometimes, trigonometry allows for unorthodox techniques providing elegant solutions which could be enlightening. Trigonometric substitutions often come into play unexpectedly. When a math technique is unexpected, it gives that technique a certain charm. This certainly adds to the beauty of mathematics.

PROBLEM 6.13.

Find x and y satisfying the inequality $x\sqrt{1-y^2} + y\sqrt{1-x^2} < 5$.

SOLUTION.

First, let's find the permissible values of both variables. This can be determined solving the following inequalities:

$$\begin{cases} 1-y^2 \geq 0, \\ 1-x^2 \geq 0; \end{cases}$$

or equivalently,

$$\begin{cases} (y-1)(y+1) \leq 0, \\ (x-1)(x+1) \leq 0. \end{cases}$$

Solution of each inequality in the system is identical and can be easily derived on a number line; any value from the segment $[-1, 1]$ satisfies each.

We see that the search for the solutions of the original inequality is restricted now to all real numbers x and y such that $|x| \leq 1$ and $|y| \leq 1$.

Recall that the range of the function sine and function cosine is all real numbers not exceeding 1 in absolute value. Therefore, for the values of x and y from the above analysis, there must exist angles α and β such that $\sin\alpha = x$ and $\sin\beta = y$. Respectively, $\sqrt{1-y^2} = \sqrt{1-\sin^2\beta} = \sqrt{\cos^2\beta} = |\cos\beta|$ and $\sqrt{1-x^2} = \sqrt{1-\sin^2\alpha} = \sqrt{\cos^2\alpha} = |\cos\alpha|$.

Let's assume that we select angles α and β such that each value under the absolute value sign is nonnegative, $|\cos\beta| = \cos\beta$ and $|\cos\alpha| = \cos\alpha$.

After substituting these expressions for x and y into the original inequality, we have on the left-hand side the sine of the sum of the angles α and β defined by the formula

$$\sin\alpha\cos\beta + \sin\beta\cos\alpha = \sin(\alpha+\beta).$$

Therefore, we can rewrite the original inequality as $\sin(\alpha+\beta) < 5$.

As we know, for any values of angles α and β, $|\sin(\alpha+\beta)| \leq 1$, because the range of the function sine is all real numbers not exceeding 1 in absolute value. It implies that any value for α and β will satisfy the inequality $\sin(\alpha+\beta) < 5$. Now, it is obvious that the values of angles α and β such that $|\cos\beta| = -\cos\beta$ and $|\cos\alpha| = -\cos\alpha$ would not affect the final outcome because we would get the sine of a sum or difference of two angles, and come to the same conclusion that the maximum value of the left-hand side of the last inequality is 1.

Turning back to our substitutions for x and y, we can now conclude that any number x and y from the interval $[-1, 1]$ will satisfy the original inequality.

Answer: $x \in [-1, 1]$ and $y \in [-1, 1]$.

What about if we are asked to solve the equation $x\sqrt{1-y^2}+y\sqrt{1-x^2}=5$?

One can save an effort trying to find the solutions, they do not exist. Clearly, based on our analysis, this equation has no real solutions!

PROBLEM 6.14.

Solve the equation $\sqrt{9-x^2}+x+\frac{2}{3}x\sqrt{9-x^2}=3$.

SOLUTION.

There are several strategies to tackle this irrational equation. For instance, we may transfer x to the right-hand side to get the equation $\sqrt{9-x^2}+\frac{2}{3}x\sqrt{9-x^2}=3-x$, and square both sides. This will transform our equation to polynomial quartic equation, not the easiest equation, though, to solve. We invite the readers to go this route to complete the solution, or you may investigate some other strategies. You should be able then to compare these options to the one we are about to explore here, and decide which one is the most efficient.

First, we will find the range of permissible values of x, i.e. the domain of our equation. In order to do it, we need to solve the inequality $9-x^2\geq0$. It can be written as

$$(x-3)(x+3)\leq0.$$

The solutions are all real numbers from the segment $[-3,3]$.

Now, let's introduce a new variable $x=3\sin\alpha$, $-\frac{\pi}{2}\leq\alpha\leq\frac{\pi}{2}$ and substitute it for x in our equation to get

$$\sqrt{9-(3\sin\alpha)^2}+3\sin\alpha+\frac{2}{3}(3\sin\alpha)\sqrt{9-(3\sin\alpha)^2}=3.$$

Note that when $-\frac{\pi}{2}\leq\alpha\leq\frac{\pi}{2}$ then $x\in[-3,3]$, so our domain restrictions are satisfied. Second, we now can use trigonometric identities $\sin^2\alpha+\cos^2\alpha=1$ and $2\sin\alpha\cos\alpha=\sin2\alpha$ to modify the new equation as

$$\sqrt{9-9\sin^2\alpha}+3\sin\alpha+2\sin\alpha\sqrt{9-9\sin^2\alpha}=3,$$
$$3\sqrt{1-\sin^2\alpha}+3\sin\alpha+6\sin\alpha\sqrt{1-\sin^2\alpha}=3,$$
$$3\sqrt{\cos^2\alpha}+3\sin\alpha+6\sin\alpha\sqrt{\cos^2\alpha}=3.$$

For $-\frac{\pi}{2}\leq\alpha\leq\frac{\pi}{2}$, $\cos\alpha\geq0$ and $\sqrt{\cos^2\alpha}=|\cos\alpha|=\cos\alpha$, so after dividing both sides by 3, we simplify the equation further to

$$\cos\alpha+\sin\alpha+2\sin\alpha\cos\alpha=1.$$

Here, we will introduce one more variable $y = \sin\alpha + \cos\alpha$. Squaring both sides yields $y^2 = \underbrace{\sin^2\alpha + \cos^2\alpha}_{1} + \underbrace{2\sin\alpha\cos\alpha}_{\sin 2\alpha} = 1 + \sin 2\alpha$, from which $\sin 2\alpha = y^2 - 1$ (*). Therefore, we obtain the new quadratic equation in y:

$y + y^2 - 1 = 1$, or equivalently, $y^2 + y - 2 = 0$. By Viète's formulas, solutions of this quadratic equation are $y = 1$ and $y = -2$. Substituting these values into (*), we get two equations to solve for α:

$$\sin 2\alpha = 1 - 1 = 0 \text{ or } \sin 2\alpha = 4 - 1 = 3.$$

The second equation has no real solutions because the range of the function sine is all real numbers not exceeding 1 in absolute value.

Solving the first equation for α restricted as $-\dfrac{\pi}{2} \le \alpha \le \dfrac{\pi}{2}$, we see that $2\alpha = 0$, from which $\alpha = 0$. Going back to our first substitution for x, we get that $x = 3\sin\alpha = 0$.

Answer: $x = 0$.

A few comments from this solution:

1. Trigonometric substitution significantly simplified the original irrational equation enabling it to be transformed into simple trigonometric equation.

2. Our second substitution $y = \sin\alpha + \cos\alpha$ and trigonometric identities $\sin^2\alpha + \cos^2\alpha = 1$ and $2\sin\alpha\cos\alpha = \sin 2\alpha$ were instrumental in tackling the trigonometric equation of a type $\cos\alpha + \sin\alpha + \sin 2\alpha = m$ (m is some real number). It is worthy to add this trigonometric substitution to your math toolbox. With this substitution, similar trigonometric equations can be solved fairly quickly.

3. The applied technique might be helpful in assessing solution strategies for similar irrational equations containing expressions x and $\sqrt{a^2 - x^2}$.

PROBLEM 6.15.

This problem was offered in GDR math Olympiad in 1980.

Solve the system of equations

$$\begin{cases} 2x + x^2 y = y, \\ 2y + y^2 z = z, \\ 2z + z^2 x = x. \end{cases}$$

SOLUTION.

Let's modify each equation in this system and rewrite it as

$$\begin{cases} y = \dfrac{2x}{1 - x^2}, \\ z = \dfrac{2y}{1 - y^2}, \\ x = \dfrac{2z}{1 - z^2}. \end{cases}$$

To solve this system of equations, we will introduce trigonometric substitution $x = \tan\alpha$, where $\alpha \in \left] -\dfrac{\pi}{2}, \dfrac{\pi}{2} \right[$. Recalling formula $\tan 2\alpha = \dfrac{2\tan\alpha}{1-\tan^2\alpha}$, we obtain that the first equation

modifies to $y = \dfrac{2\tan\alpha}{1-\tan^2\alpha} = \tan 2\alpha$, the second equation modifies to $z = \dfrac{2\tan 2\alpha}{1-\tan^2 2\alpha} = \tan 4\alpha$,

and the third equation modifies to $x = \dfrac{2\tan 4\alpha}{1-\tan^2 4\alpha} = \tan 8\alpha$.

Therefore, we can now find x solving the equation $\tan\alpha = \tan 8\alpha$. This can be rewritten

as $\dfrac{\sin 8\alpha}{\cos 8\alpha} - \dfrac{\sin\alpha}{\cos\alpha} = 0$. After several simplifications, recalling the trigonometric identity

$\sin(\beta - \gamma) = \sin\beta \cdot \cos\gamma - \sin\gamma \cdot \cos\beta$, we have

$$\frac{\sin 8\alpha \cdot \cos\alpha - \sin\alpha \cdot \cos 8\alpha}{\cos 8\alpha \cdot \cos\alpha} = 0,$$

$$\sin 7\alpha = 0, \ (\cos 8\alpha \neq 0, \ \cos\alpha \neq 0),$$

$$\alpha = \frac{\pi n}{7}, \ n \epsilon Z.$$

Therefore, $x = \tan\alpha = \tan\left(\dfrac{\pi n}{7}\right)$, $y = \tan 2\alpha = \tan\left(\dfrac{2\pi n}{7}\right)$, and $z = \tan 4\alpha = \tan\left(\dfrac{4\pi n}{7}\right)$. We

suggest that readers make the final step in this solution independently and verify that for $n = \pm 1$, $n = 0$, $n = \pm 2$, and $n = \pm 3$ there will be 7 distinct sets of the solutions for x, y, and z satisfying the original system of equations.

PROBLEM 6.16.
Find the maximum and the minimum of the function $y = \dfrac{1+x^4}{\left(1+x^2\right)^2}$ without resorting to calculus means.

SOLUTION.
As many problems in our book, this one has alternative solutions. Once again, we encourage the readers to investigate different approaches to its solution. We will investigate here

trigonometric substitution $x = \tan\varphi$, $-\dfrac{\pi}{2} < \varphi < \dfrac{\pi}{2}$. Observing that the domain of the given

function is all real numbers, $x \in R$, the suggested substitution $x = \tan\varphi$ is valid for the

selected values of φ, $-\dfrac{\pi}{2} < \varphi < \dfrac{\pi}{2}$.

This allows us to modify our function to its trigonometric equivalent presentation, which significantly simplifies the path to the desired result.

Recalling that $\sin^2\varphi + \cos^2\varphi = 1$ and $2\sin\varphi \cdot \cos\varphi = \sin 2\varphi$ gives

$$y = \frac{1+\tan^4\varphi}{\left(1+\tan^2\varphi\right)^2} = \frac{1+\dfrac{\sin^4\varphi}{\cos^4\varphi}}{\left(1+\dfrac{\sin^2\varphi}{\cos^2\varphi}\right)^2} = \frac{\dfrac{\cos^4\varphi + \sin^4\varphi}{\cos^4\varphi}}{\dfrac{\left(\cos^2\varphi + \sin^2\varphi\right)^2}{\cos^4\varphi}} = \cos^4\varphi + \sin^4\varphi =$$

$$(\cos^4\varphi + 2\sin^2\varphi \cdot \cos^2\varphi + \sin^4\varphi) - 2\sin^2\varphi \cdot \cos^2\varphi = \left(\sin^2\varphi + \cos^2\varphi\right)^2 - \frac{1}{2}\sin^2 2\varphi =$$

$$1 - \frac{1}{2}\sin^2 2\varphi.$$

So, we see that $y = 1 - \dfrac{1}{2}\sin^2 2\varphi$, $-\dfrac{\pi}{2} < \varphi < \dfrac{\pi}{2}$. Because $0 \le \sin^2 2\varphi \le 1$, we conclude that

$\dfrac{1}{2} \le 1 - \dfrac{1}{2}\sin^2 2\varphi \le 1$, i.e. $\dfrac{1}{2} \le y \le 1$. Therefore, the minimum of our function equals $\dfrac{1}{2}$ and

it is attained for $\varphi = -\dfrac{\pi}{4}$ and for $\varphi = \dfrac{\pi}{4}$, whilst the maximum equals 1 and it is attained

for $\varphi = 0$. This implies that the minimum of the original function is $\dfrac{1}{2}$ and it is attained

at $x = \tan\left(-\dfrac{\pi}{4}\right) = -1$ and $x = \tan\dfrac{\pi}{4} = 1$; the maximum of the original function is 1 and it is

attained at $x = 0$, which are the answers to the problem.

PROBLEM 6.17.

Find all real values of a such that the equation $x\sqrt{1-4x^2} \cdot (1-8x^2) = a$ has solutions.

SOLUTION.

First, we need to determine the permissible values of x, i.e. find the domain of the expression in the left-hand side.

We need to solve inequality $1 - 4x^2 \ge 0$. Factoring $1 - 4x^2$ as the difference of squares and multiplying both sides of inequality by $-\dfrac{1}{4}$, gives

$$(1-2x)(1+2x) \ge 0,$$

$$\left(x - \dfrac{1}{2}\right)\left(x + \dfrac{1}{2}\right) \le 0.$$

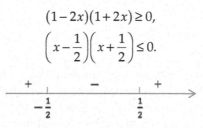

We see that $x \in \left[-\dfrac{1}{2}, \dfrac{1}{2}\right]$. So, since $|x| \le 1$, we can try to use trigonometric substitution

$x = \dfrac{1}{2}\sin\varphi$, $\varphi \in \left[-\dfrac{\pi}{2}, \dfrac{\pi}{2}\right]$. Our equation then is modified to

$$\dfrac{1}{2}\sin\varphi \cdot \sqrt{1-\sin^2\varphi} \cdot (1-2\sin^2\varphi) = a.$$

Recalling trigonometric identities $\sin^2\varphi + \cos^2\varphi = 1$ and $\cos^2\varphi - \sin^2\varphi = \cos 2\varphi$, we can

further modify the equation to $\dfrac{1}{2}\sin\varphi \cdot \sqrt{\cos^2\varphi} \cdot \cos 2\varphi = a$. Observing that for $\varphi \in \left[-\dfrac{\pi}{2}, \dfrac{\pi}{2}\right]$,

$|\cos\varphi| = \cos\varphi$, we have $\dfrac{1}{2}\sin\varphi \cdot \cos\varphi \cdot \cos 2\varphi = a$.

Finally, multiplying both sides of the last equation by 8 and applying twice the formula for the sine of a double angle, $2\sin\alpha \cdot \cos\alpha = \sin 2\alpha$, we arrive at the equation

$$\sin 4\varphi = 8a.$$

Range of the function sine is all real numbers not exceeding 1 in absolute value. Therefore, the last equation will have solutions when $-\dfrac{1}{8} \le a \le \dfrac{1}{8}$.

Answer: $a \epsilon \left[-\dfrac{1}{8}, \dfrac{1}{8} \right]$.

Trigonometric substitutions broadly come into play in calculus in finding the area of a circle or an ellipse; an integral of the form $\int \sqrt{a^2 - x^2}\, dx$ arises, where $a > 0$. As we just evidenced in our solution above, when we change the variable from x to β by substitution $x = a \sin \beta$, then the trigonometric identity $1 - \sin^2 \beta = \cos^2 \beta$ allows us to get rid of irrationality because $\sqrt{a^2 - x^2} = \sqrt{a^2 - a^2 \sin^2 \beta} = a\sqrt{1 - \sin^2 \beta} = a|\cos \beta|$. It is important to remember that we can make trigonometric substitution provided that it defines a one-to-one function. This can be accomplished by assessing the restrictions on β that $-\dfrac{\pi}{2} \le \beta \le \dfrac{\pi}{2}$.

The following table provides the list of useful trigonometric substitutions. Each one can be applied only for specific restrictions imposed on argument ensuring that the function that defines the substitution is one-to-one:

Expression	Substitution	Identity
$\sqrt{a^2 - x^2}$	$x = a \sin \beta,\ -\dfrac{\pi}{2} \le \beta \le \dfrac{\pi}{2}$	$1 - \sin^2 \beta = \cos^2 \beta$
$\sqrt{a^2 + x^2}$	$x = a \tan \beta,\ -\dfrac{\pi}{2} < \beta < \dfrac{\pi}{2}$	$1 + \tan^2 \beta = \sec^2 \beta$
$\sqrt{x^2 - a^2}$	$x = a \sec \beta,\ 0 \le \beta < \dfrac{\pi}{2}$ or $\pi \le \beta < \dfrac{3\pi}{2}$	$\sec^2 \beta - 1 = \tan^2 \beta$

Our discussion about utilizing trigonometry in problem solving would not be complete without mentioning the name of the prominent French mathematician François Viète (1540–1603). He was the founder of symbolic algebra, being the first to introduce the idea for representing known and unknown quantities by letters (variables); and that is why he is often regarded as the father of new algebra. His studies tremendously contributed in developing algebra, trigonometry, geometry, and astronomy. He was also one of the first mathematicians who efficiently utilized trigonometry as connecting link between algebraic and geometrical issues.

He developed an interesting technique for the trigonometric solution for the cubic equation of the form $x^3 + px + q = 0$ (p and q are some real numbers) with three real roots.

Viète was solving a geometric problem of the section of an angle into an odd number of equal parts. He described how the chord AB of a trisected arc associated with chord BD (see figure below) of a circle of a given radius r, is a solution of a cubic equation.

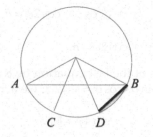

His name is also associated with solving the very interesting equation suggested by the Flemish mathematician Adriaan Van Roomen (1561–1615) in 1593.

This frightening equation is

$$x^{45} - 45x^{43} + 945x^{41} - 12300x^{39} + 111150x^{37} - 740259x^{35} + 3764565x^{33} - 14945040x^{31}$$
$$+ 46955700x^{29} - 117679100x^{27} + 236030652x^{25} - 378658800x^{23} + 483841800x^{21}$$
$$- 488494125x^{19} + 384942375x^{17} - 232676280x^{15} + 105306075x^{13} - 34512075x^{11}$$
$$+ 7811375x^{9} - 1138500x^{7} + 95634x^{5} - 3795x^{3} + 45x = \sqrt{\frac{7}{4} - \sqrt{\frac{5}{16}} - \sqrt{\frac{15}{8} - \sqrt{\frac{45}{64}}}}.$$

François Viète solved it by applying the substitution $x = 2\sin\alpha$.

We are not going to cover in details the solutions of these two problems here. The ambitious readers can find more about these interesting discoveries in mathematical literature; for instance, both were mentioned in B. Pritsker, "The Equations World", Dover Publications, 2019.

7

Euclidean Vectors

Where there is matter, there is geometry.

Johannes Kepler

A Euclidean vector is a geometric object defined by its length and direction. Vectors play an important role, not only in mathematics, but in other disciplines, such as physics and engineering. Many physical quantities can be thought of as vectors, for example, the velocity and acceleration of a moving object and the forces acting on it, displacement, linear acceleration, angular acceleration, linear momentum, and angular momentum can be all described with vectors. In mathematics, properties of Euclidean vectors are very important in establishing connections between algebraic and geometric issues and they are exceptionally useful in solving many problems, providing a great alternative to conventional techniques. In this chapter, we will provide a broad presentation of the beauty of problem-solving through the application of Euclidean vector properties.

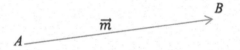

Vector AB is denoted as $\overrightarrow{AB} = \vec{m}$. Its length or magnitude is the length of the segment AB and it has a direction from A to B. The direction of the vector is denoted by the arrow at the terminal point. Two vectors are equal if they have the same length and direction. Operations with vectors are analogous to algebraic operations on real numbers of addition, subtraction, multiplication, and negation. These operations also comply with the algebraic laws of commutativity, associativity, and distributivity. The addition of two vectors may be represented geometrically by the *triangle rule* or by the *parallelogram rule*.

Triangle rule:

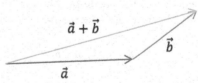

We place the initial point of the vector $\vec{b}$ at the terminal point of the vector $\vec{a}$, and then we draw a vector from the initial point of $\vec{a}$ to the terminal point of $\vec{b}$. The newly formed vector $\vec{a} + \vec{b}$ represents the sum of the vectors $\vec{a}$ and $\vec{b}$.

DOI: 10.1201/9781003359500-7

Parallelogram rule:

When we add two vectors $\vec{a}$ and $\vec{b}$ that have the common initial point (as it is shown in the figure below), we construct the parallelogram on these vectors as on the sides, and draw the diagonal from their common initial point.

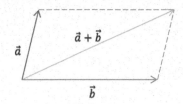

The vector that has this common initial point and the terminal point at the opposite vertex of the parallelogram is the sum of the vectors $\vec{a}$ and $\vec{b}$, the vector $\vec{a}+\vec{b}$.

Two important vector equalities easily evolve from the above-mentioned definition.

Lemma. For any random point M outside of a parallelogram $ABCD$ and point O, the intersection of its diagonals, the following vector equalities hold:

$$\overrightarrow{MO} = \frac{1}{2}\left(\overrightarrow{MB}+\overrightarrow{MD}\right) = \frac{1}{2}\left(\overrightarrow{MA}+\overrightarrow{MC}\right).$$

Indeed, it suffices to extend MO to get point M' such that $MO = OM'$. Connecting M' with A, B, C, and D two parallelograms $AMCM'$ and $BMDM'$ will be formed. Applying the parallelogram rule we can now easily get the above vector equalities.

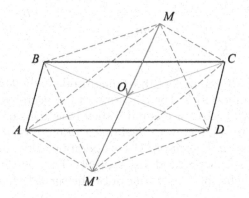

Subtraction of two vectors can be geometrically illustrated as follows: to subtract $\vec{b}$ from $\vec{a}$, place the initial point of both vectors at the same point, and then draw the vector from the terminal point of $\vec{b}$ to the terminal point of $\vec{a}$ (see figure below).

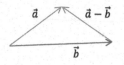

A vector may be multiplied by a real number (or scalar) k. The operation of multiplying a vector by a scalar is called scalar multiplication. Multiplying by a scalar k stretches a vector out by a factor of k. If k is a negative number, the resulting vector has an opposite direction.

For example, by multiplying vector $\vec{a}$ by 3, we get the resulting vector with the same direction and with the length 3 times of the vector $\vec{a}$:

By multiplying vector $\vec{a}$ by –2 we get the vector with the opposite direction and with the length twice of the vector $\vec{a}$:

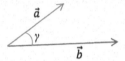

Scalar multiplication of a vector by a number should not be confused with the scalar multiplication of two vectors. The *dot product* (or a *scalar product*) of two non-zero vectors $\vec{a}$ and $\vec{b}$ is defined as the product of the length of $\vec{a}$ by the length of $\vec{b}$ and by the cosine of the angle between them: $\vec{a} \cdot \vec{b} = \|\vec{a}\| \cdot \|\vec{b}\| \cdot \cos \gamma$, where $\|\vec{a}\|$ is the length of the vector $\vec{a}$, $\|\vec{b}\|$ is the length of the vector $\vec{b}$, and γ is the angle these vectors form.

The following important property evolves from the above definition of the dot product of two vectors: *two non-zero vectors are perpendicular if and only if their dot product equals to 0.*

Direct statement:

Given: $\vec{a} \perp \vec{b}$, $\|\vec{a}\| \neq 0$ and $\|\vec{b}\| \neq 0$.
To Prove: $\vec{a} \cdot \vec{b} = 0$.

Proof.
To prove the direct statement, we merely have to observe that since $\vec{a} \perp \vec{b}$ then the angle between these two vectors is 90°. Therefore,

$$\vec{a} \cdot \vec{b} = \|\vec{a}\| \cdot \|\vec{b}\| \cdot \cos 90° = \|\vec{a}\| \cdot \|\vec{b}\| \cdot 0 = 0$$

Converse statement:

Given: $\vec{a} \cdot \vec{b} = 0$, $\|\vec{a}\| \neq 0$ and $\|\vec{b}\| \neq 0$.
To Prove: $\vec{a} \perp \vec{b}$.

Proof.
Dot product of two vectors is defined as $\vec{a} \cdot \vec{b} = \|\vec{a}\| \cdot \|\vec{b}\| \cdot \cos \gamma$.

It is known that $\|\vec{a}\| \neq 0$ and $\|\vec{b}\| \neq 0$. Hence, the dot product of the vectors $\vec{a}$ and $\vec{b}$ will equal to 0 only when $\cos \gamma = 0$. This implies that $\gamma = 90°$, i.e., $\vec{a} \perp \vec{b}$, which is what we wished to prove.

PROBLEM 7.1.

Let $ABCD$ be a convex quadrilateral in which the sums of the squares of the opposite sides are equal to each other, $AB^2 + CD^2 = BC^2 + AD^2$. Prove that the diagonals of $ABCD$ are perpendicular, $AC \perp BD$.

<u>PROOF.</u>

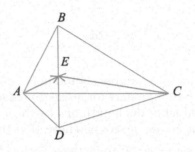

Denote E the midpoint of BD. Applying the parallelogram rule, we have $2\vec{AE} = \vec{AB} + \vec{AD}$. On the other hand, $\vec{BD} = \vec{AD} - \vec{AB}$. Now, we will find the dot product of vectors $2\vec{AE}$ and $\vec{BD}$:

$$2\vec{AE} \cdot \vec{BD} = (\vec{AB} + \vec{AD}) \cdot (\vec{AD} - \vec{AB}) = \vec{AB} \cdot \vec{AD} - AB^2 + AD^2 - \vec{AD} \cdot \vec{AB} = AD^2 - AB^2.$$

Hence,

$$2\vec{AE} \cdot \vec{BD} = AD^2 - AB^2. \tag{7.1}$$

Analogously, $2\vec{CE} = \vec{CB} + \vec{CD}$ and $\vec{BD} = \vec{CD} - \vec{CB}$. The dot product of vectors $2\vec{CE}$ and $\vec{BD}$ is

$$2\vec{CE} \cdot \vec{BD} = (\vec{CB} + \vec{CD}) \cdot (\vec{CD} - \vec{CB}) = \vec{CB} \cdot \vec{CD} - CB^2 + CD^2 - \vec{CD} \cdot \vec{CB} = CD^2 - CB^2.$$

Hence,

$$2\vec{CE} \cdot \vec{BD} = CD^2 - CB^2. \tag{7.2}$$

Subtracting (7.2) from (7.1) and recalling that $AB^2 + CD^2 = BC^2 + AD^2$, gives

$$2\vec{AE} \cdot \vec{BD} - 2\vec{CE} \cdot \vec{BD} = AD^2 - AB^2 - CD^2 + CB^2 = (AD^2 + CB^2) - (AB^2 + CD^2) = 0.$$

Thus, $2\vec{AE} \cdot \vec{BD} - 2\vec{CE} \cdot \vec{BD} = 0$, or equivalently, $2\vec{BD} \cdot (\vec{AE} - \vec{CE}) = 0$. Since $\vec{AE} - \vec{CE} = \vec{AC}$, the last equality can be rewritten as $\vec{BD} \cdot \vec{AC} = 0$. It implies that $AC \perp BD$, as it was required to be proved.

Analyzing the problem we just solved, it is not hard to come up with the so-called *Rule of four points* which states that for any four points A, B, C, and D,

$$\vec{DA} \cdot \vec{BC} + \vec{DB} \cdot \vec{CA} + \vec{DC} \cdot \vec{AB} = 0.$$

Let's prove it.

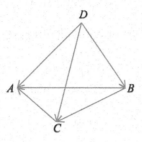

Consider the following differences of two vectors:

$$\overrightarrow{BC} = \overrightarrow{DC} - \overrightarrow{DB},$$
$$\overrightarrow{CA} = \overrightarrow{DA} - \overrightarrow{DC},$$
$$\overrightarrow{AB} = \overrightarrow{DB} - \overrightarrow{DA}.$$

Now, multiplying each equality respectively by $\overrightarrow{DA}, \overrightarrow{DB}$, and $\overrightarrow{DC}$, and adding three of them with further cancelling out like terms yields

$$\overrightarrow{DA} \cdot \overrightarrow{BC} + \overrightarrow{DB} \cdot \overrightarrow{CA} + \overrightarrow{DC} \cdot \overrightarrow{AB} = \overrightarrow{DA}\left(\overrightarrow{DC} - \overrightarrow{DB}\right) + \overrightarrow{DB}\left(\overrightarrow{DA} - \overrightarrow{DC}\right) + \overrightarrow{DC}\left(\overrightarrow{DB} - \overrightarrow{DA}\right) =$$
$$\overrightarrow{DA} \cdot \overrightarrow{DC} - \overrightarrow{DA} \cdot \overrightarrow{DB} + \overrightarrow{DB} \cdot \overrightarrow{DA} - \overrightarrow{DB} \cdot \overrightarrow{DC} + \overrightarrow{DC} \cdot \overrightarrow{DB} - \overrightarrow{DC} \cdot \overrightarrow{DA} = 0.$$

Indeed, we arrive at $\overrightarrow{DA} \cdot \overrightarrow{BC} + \overrightarrow{DB} \cdot \overrightarrow{CA} + \overrightarrow{DC} \cdot \overrightarrow{AB} = 0$, which is the desired result.

The ideas behind the method considered above prove useful in tackling three-dimensional geometric problems as well.

PROBLEM 7.2.

A tetrahedron $ABCD$ is orthocentric if and only if the sum of the squares of opposite edges is the same for the three pairs of opposite edges.

SOLUTION.

First, recall that by definition, an orthocentric tetrahedron is a tetrahedron where all three pairs of opposite edges are perpendicular.

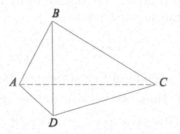

We will start with the direct statement and prove that if $ABCD$ is an orthocentric tetrahedron, then $AB^2 + CD^2 = BC^2 + DA^2$.

It is not hard to prove that $\overrightarrow{AB} + \overrightarrow{BC} + \overrightarrow{CD} + \overrightarrow{DA} = 0$. Indeed, $\overrightarrow{AB} + \overrightarrow{BC} = \overrightarrow{AC}$. Next, $\overrightarrow{AC} + \overrightarrow{CD} = \overrightarrow{AD}$, and finally, $\overrightarrow{AD} + \overrightarrow{DA} = 0$. Using this vector equality we get that

$$\overrightarrow{AB} + \overrightarrow{BC} = -\overrightarrow{CD} - \overrightarrow{DA}, \text{ or equivalently, } \overrightarrow{AB} + \overrightarrow{CD} = -\overrightarrow{BC} - \overrightarrow{DA}.$$

Squaring the last equality gives $\left(\overrightarrow{AB} + \overrightarrow{CD}\right)^2 = \left(-\overrightarrow{BC} - \overrightarrow{DA}\right)^2$. Therefore, recalling that the opposite edges are perpendicular, i.e., $\overrightarrow{AB} \cdot \overrightarrow{CD} = \overrightarrow{BC} \cdot \overrightarrow{DA} = 0$, we obtain that

$$AB^2 + 2\underbrace{\overrightarrow{AB} \cdot \overrightarrow{CD}}_{0} + CD^2 = BC^2 + 2\underbrace{\overrightarrow{BC} \cdot \overrightarrow{DA}}_{0} + DA^2,$$

which yields the sought-after result, $AB^2 + CD^2 = BC^2 + DA^2$. The proof of the equality for any other pairs of opposite edges is done in the same manner as that shown above.

To prove the converse statement, we can follow the solution of Problem 7.1. The beauty of the vector algebra technique is that it is equally effective in both, the two-dimensional plane and three-dimensional space. Even though we were working with the convex quadrilateral in Problem 7.1, exactly the same way of thinking "vectors-wise" will be applicable to the tetrahedron.

PROBLEM 7.3.

Let point O be the centroid of a triangle ABC. Prove that

$$\overrightarrow{OA} + \overrightarrow{OB} + \overrightarrow{OC} = 0.$$

SOLUTION.

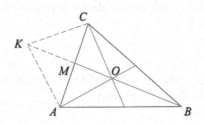

Let's do the auxiliary construction and extend BM (M is the mid-point of AC) to locate point K on the straight line BM such that $KM = OM$. Connecting A and K and C and K we obtain parallelogram $AKCO$. It is a parallelogram because by our constructions, the point of intersection of its diagonals M divides the diagonals in half, $AM = CM$ and $KM = OM$. Therefore, by applying the parallelogram rule,

$$\overrightarrow{OA} + \overrightarrow{OC} = \overrightarrow{OK}. \tag{7.3}$$

The centroid of a triangle is the point of intersection of its medians and it divides each of the medians in the ratio 2:1. Thus, $OM = \frac{1}{2}OB$, and we can state that $\overrightarrow{OK} = 2\overrightarrow{OM} = -\overrightarrow{OB}$ (the length of the vector $\overrightarrow{OK}$ is the same as the length of the vector $\overrightarrow{OB}$ and they have the opposite directions). It follows that $\overrightarrow{OK} + \overrightarrow{OB} = 0$. Now, substituting the expression for $\overrightarrow{OK}$ from (7.3) into the last equality yields $\overrightarrow{OA} + \overrightarrow{OC} + \overrightarrow{OB} = 0$, which is the vector equality that we were required to prove.

PROBLEM 7.4.

Given a triangle with the sides a, b, and c, and its circumradius R, prove that

$$a^2 + b^2 + c^2 \le 9R^2.$$

SOLUTION.

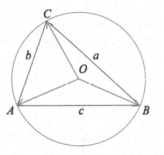

We have triangle ABC inscribed in the circle with center O and radius $R = OA = OB = OC$. We use the traditional notation for the sides of the triangle: $BC = a$, $AC = b$, and $AB = c$. To prove the requested inequality, we will apply the vectors properties and express each of the vectors $\overrightarrow{OA}$, $\overrightarrow{OB}$, and $\overrightarrow{OC}$ as the difference of two vectors:

$$\vec{a} = \overrightarrow{CB} = \overrightarrow{OB} - \overrightarrow{OC},$$
$$\vec{b} = \overrightarrow{CA} = \overrightarrow{OA} - \overrightarrow{OC},$$
$$\vec{c} = \overrightarrow{AB} = \overrightarrow{OB} - \overrightarrow{OA}.$$

Squaring the above equalities and adding them yields (keeping in mind that for any vector $\vec{m}$, $\vec{m}^2 = \|\vec{m}\|^2 = m^2$):

$$a^2 + b^2 + c^2 = \left(\overrightarrow{OB} - \overrightarrow{OC}\right)^2 + \left(\overrightarrow{OA} - \overrightarrow{OC}\right)^2 + \left(\overrightarrow{OB} - \overrightarrow{OA}\right)^2 =$$

$$OB^2 + OC^2 - 2\overrightarrow{OB} \cdot \overrightarrow{OC} + OA^2 + OC^2 - 2\overrightarrow{OA} \cdot \overrightarrow{OC} + OB^2 + OA^2 - 2\overrightarrow{OB} \cdot \overrightarrow{OA} =$$

$$2\left(OB^2 + OC^2 + OA^2\right) - 2\left(\overrightarrow{OB} \cdot \overrightarrow{OC} + \overrightarrow{OA} \cdot \overrightarrow{OC} + \overrightarrow{OB} \cdot \overrightarrow{OA}\right). \qquad (7.4)$$

Now, we recall that

$$R = OA = OB = OC. \qquad (7.5)$$

Also, it's not hard to see (squaring in the right side and simplifying suffices) that

$$2\left(\overrightarrow{OB} \cdot \overrightarrow{OC} + \overrightarrow{OA} \cdot \overrightarrow{OC} + \overrightarrow{OB} \cdot \overrightarrow{OA}\right) = \left(\overrightarrow{OB} + \overrightarrow{OC} + \overrightarrow{OA}\right)^2 - \left(OB^2 + OC^2 + OA^2\right). \qquad (7.6)$$

Substituting the expressions from (7.5) and (7.6) into (7.4) gives

$$a^2 + b^2 + c^2 = 2\left(R^2 + R^2 + R^2\right) - \left(\left(\overrightarrow{OB} + \overrightarrow{OC} + \overrightarrow{OA}\right)^2 - \left(R^2 + R^2 + R^2\right)\right) =$$

$$6R^2 - \left(\overrightarrow{OB} + \overrightarrow{OC} + \overrightarrow{OA}\right)^2 + 3R^2 = 9R^2 - \underbrace{\left(\overrightarrow{OB} + \overrightarrow{OC} + \overrightarrow{OA}\right)^2}_{\ge 0} \le 9R^2.$$

So, after our manipulations we see that indeed, $a^2 + b^2 + c^2 \leq 9R^2$, as it was requested to be proved. Moreover, based on the last step in our proof, we may conclude that the equality is attained only when $\overrightarrow{OB} + \overrightarrow{OC} + \overrightarrow{OA} = 0$. Referring to the previous problem, this equality is possible only when the circumcenter coincides with the triangle's centroid, which implies that such a triangle has to be an equilateral triangle.

PROBLEM 7.5.

There is given an equilateral triangle with side a. Prove that the sum of the squares of the distances from any arbitrary point on the triangle's circumcircle to its vertices equals $2a^2$.

SOLUTION.

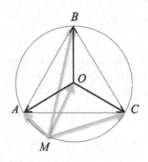

We are given the equilateral triangle ABC ($AB = AC = BC = a$) inscribed in the circle with center O. Let M be some randomly selected point on the circumcircle of $\triangle ABC$. The goal is to prove that $MA + MB + MC = 2a^2$.

Expressing each of the vectors $\overrightarrow{MA}$, $\overrightarrow{MB}$, and $\overrightarrow{MC}$ as the sum of two vectors, we have

$$\overrightarrow{MA} = \overrightarrow{MO} + \overrightarrow{OA},$$
$$\overrightarrow{MB} = \overrightarrow{MO} + \overrightarrow{OB},$$
$$\overrightarrow{MC} = \overrightarrow{MO} + \overrightarrow{OC}.$$

Observing that $MO = OA = OB = OC = R$, where R is the radius of our circumcircle, squaring each of the above three vector equalities, and adding yields

$$MA^2 + MB^2 + MC^2 = MO^2 + 2\overrightarrow{MO} \cdot \overrightarrow{OA} + OA^2 + MO^2 + 2\overrightarrow{MO} \cdot \overrightarrow{OB} + OB^2 + MO^2 +$$
$$2\overrightarrow{MO} \cdot \overrightarrow{OC} + OC^2 = 6R^2 + 2\overrightarrow{MO} \cdot \left(\overrightarrow{OA} + \overrightarrow{OB} + \overrightarrow{OC}\right) (*)$$

In an equilateral triangle the center of its circumcircle is the centroid as well, so applying the result from Problem 7.3, we have $\overrightarrow{OA} + \overrightarrow{OB} + \overrightarrow{OC} = 0$. Also, it's easy to express the circumradius of the equilateral triangle in terms of its side as $R = \dfrac{a}{\sqrt{3}}$ (we hope that readers can easily derive this formula, or they can find the proof in the appendix). Substituting these values into (*) gives the sought-after result

$$MA^2 + MB^2 + MC^2 = 6 \cdot \frac{a^2}{3} + 0 = 2a^2.$$

It is noteworthy that this solution can be extended for any regular *n*-gon, not necessarily circumscribed, but inscribed as· well. This also holds true for 3-dimentional space. For instance, the suggested solution can be applied to the following problem:

There is given a regular triangle with the side *a*, and its three vertices lie on the sphere, the center of which coincides with the center of the given triangle. Prove that the sum of the squares of the distances from any arbitrary point on the sphere to the triangle's vertices equals $2a^2$.

It should be a good exercise for the readers to reproduce these proofs by themselves.

PROBLEM 7.6.

Given that *O* is the circumcenter (center of the circumscribed circle) and *H* is the orthocenter (point of intersection of altitudes) of a triangle *ABC*, prove that

$$\overrightarrow{OH} = \overrightarrow{OA} + \overrightarrow{OB} + \overrightarrow{OC}.$$

<u>PROOF.</u>

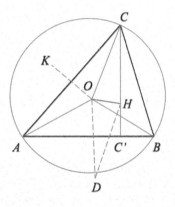

First, we will construct *D*, the image of point *O* in reflection through *AB* axis. Then by the parallelogram rule, $\overrightarrow{OD} = \overrightarrow{OA} + \overrightarrow{OB}$. Next, we will locate *H* such that $\overrightarrow{OH} = \overrightarrow{OD} + \overrightarrow{OC}$. Substituting expression for $\overrightarrow{OD}$ into the last equality, we can rewrite it as

$$\overrightarrow{OH} = \overrightarrow{OD} + \overrightarrow{OC} = \overrightarrow{OA} + \overrightarrow{OB} + \overrightarrow{OC}.$$

Now, it remains to show that such identified point *H* is the orthocenter of △*ABC*.

Note that to find *H* such that $\overrightarrow{OH} = \overrightarrow{OD} + \overrightarrow{OC}$, we need to apply the parallelogram rule. So, *H* should be such that $CH \parallel OD$ and $CH = OD$. Since *D* is the image of *O* in reflection through *AB* axis, then $OD \perp AB$. It follows that being parallel to *OD*, *CH* is also perpendicular to *AB*, i.e., *H* belongs to the altitude dropped from vertex *C* to *AB*. If we repeat our constructions, starting now from vectors $\overrightarrow{OA}$ and $\overrightarrow{OC}$, we will get the same point *H*. But now it will be lying on the altitude dropped from *B* to *AC*. Indeed, constructing *K*, the image of *O* in reflection through *AC* axis, we get that

$$\overrightarrow{OK} = \overrightarrow{OA} + \overrightarrow{OC}.$$

Now, finding the sum of vectors $\overrightarrow{OK}$ and $\overrightarrow{OB}$, we will find some vector, let's say $\overrightarrow{OH'}$, such that $\overrightarrow{OH'} = \overrightarrow{OK} + \overrightarrow{OB} = \overrightarrow{OA} + \overrightarrow{OB} + \overrightarrow{OC}$. The last equality implies that H' and H coincide.

In a similar fashion we can get that H belongs to the third altitude of the triangle dropped from A to BC. In passing, we just proved the fact that the orthocenter of a triangle, the point of intersection of all three altitudes, indeed exists, and as shown above, this point H is such that $\overrightarrow{OH} = \overrightarrow{OA} + \overrightarrow{OB} + \overrightarrow{OC}$. This is the result we need.

The vector equality studied in Problem 7.6 proves useful in many problems concerning the orthocenter of a triangle. We will demonstrate this in Problems 7.7 and 7.8.

PROBLEM 7.7.

Prove that points symmetrical to the orthocenter of a triangle with respect to the midpoints of its sides lie on its circumcircle.

PROOF.

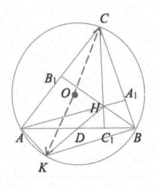

Let H be the orthocenter of the triangle ABC, O is the center of its circumcircle, and D is the midpoint of AB. Construct K symmetrical to H with respect to D. Then $DK = DH$. We see that two segments AB and HK have the common midpoint D. Therefore, $AHBK$ is a parallelogram (its diagonals AB and HK bisect each other).

As we observed at the beginning of this chapter, by Lemma, using the parallelogram rule, it follows that $\overrightarrow{OD} = \frac{1}{2}\left(\overrightarrow{OA} + \overrightarrow{OB}\right)$ and $\overrightarrow{OD} = \frac{1}{2}\left(\overrightarrow{OH} + \overrightarrow{OK}\right)$. Equating the right sides of these equalities, we get that

$$\overrightarrow{OA} + \overrightarrow{OB} = \overrightarrow{OH} + \overrightarrow{OK}. \tag{7.7}$$

As it was proved in the previous problem, if O is the center of the circumcircle of the triangle ABC, then $\overrightarrow{OH} = \overrightarrow{OA} + \overrightarrow{OB} + \overrightarrow{OC}$. Substituting the expression for $\overrightarrow{OH}$ from this equality into (7.7) yields $\overrightarrow{OA} + \overrightarrow{OB} = \overrightarrow{OA} + \overrightarrow{OB} + \overrightarrow{OC} + \overrightarrow{OK}$, from which we derive that $\overrightarrow{OC} = -\overrightarrow{OK}$. "Translating" this vector equality into geometrical terms, we get that $\overrightarrow{OC}$ and $\overrightarrow{OK}$ are of the same length and have the opposite directions. This implies that CK is the diameter of the circumcircle of the triangle ABC, or, in other words, K lies on the circumcircle of the triangle ABC. Repeating similar observations for the other two sides BC and AC, we will find that images of H with respect to the midpoints of these sides lie on the circumcircle as well, completing the proof of the problem's assertion.

PROBLEM 7.8.

Prove that $AH = 2r \cdot |\cos A|$ if H is the orthocenter of the triangle ABC and r is the radius of its circumcircle.

PROOF.

We will use the same figure as in the previous problem. According to Problem 7.6,

$$\overrightarrow{OH} = \overrightarrow{OA} + \overrightarrow{OB} + \overrightarrow{OC}.$$

Applying the triangle rule, we get that $\overrightarrow{OH} = \overrightarrow{OA} + \overrightarrow{AH}$, and substituting this expression in the above equality yields $\overrightarrow{OA} + \overrightarrow{AH} = \overrightarrow{OA} + \overrightarrow{OB} + \overrightarrow{OC}$. It follows that $\overrightarrow{AH} = \overrightarrow{OB} + \overrightarrow{OC}$.

Hence, $\overrightarrow{AH}^2 = \left(\overrightarrow{OB} + \overrightarrow{OC}\right)^2$.

$$\overrightarrow{AH}^2 = \left\|\overrightarrow{OB}^2\right\| + \left\|\overrightarrow{OC}^2\right\| + 2\left\|\overrightarrow{OB}\right\| \cdot \left\|\overrightarrow{OC}\right\| \cdot \cos \angle BOC = 2r^2 + 2r^2 \cdot \cos \angle BOC = 2r^2 \left(1 + \cos \angle BOC\right) \;(*).$$

Angle BOC is the central angle subtended by the chord BC. According to the Inscribed Angle Theorem, the measure of the central angle $\angle BOC$ equals twice the measure of the inscribed angle $\angle BAC$, that is, $\angle BOC = 2\angle BAC = 2\angle A$. Substituting this into (*) and engaging some trigonometry, recalling that $1 + \cos 2A = \sin^2 A + \cos^2 A + \cos^2 A - \sin^2 A = 2\cos^2 A$, we obtain that $\overrightarrow{AH}^2 = 2r^2 \left(1 + \cos \angle BOC\right) = 2r^2 \cdot 2\cos^2 A = \left(2r \cdot \cos A\right)^2$. Therefore, we finally arrive at the desired equality $AH = 2r \cdot |\cos A|$.

In the Cartesian coordinate system (the detail discussion to follow in the next chapter), a vector is represented by identifying the coordinates of its initial and terminal point, and in two-dimensional plane it is written as $\vec{a} = (x, y)$. For example, the vector from the origin O $(0, 0)$ to the point A $(4, 1)$ is written as $\vec{a} = \overrightarrow{OA} = (4, 1)$ (see figure below).

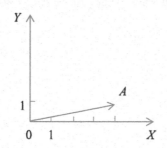

To add or subtract vectors, we add and subtract their respective coordinates to get the coordinates of the resulting vector, their sum or difference. To find the dot product of two vectors represented by the coordinates, $\vec{a} = (x, y)$ and $\vec{b} = (m, n)$, the following formula is used: $\vec{a} \cdot \vec{b} = xm + yn$. For collinear vectors (they lie on the same line or on parallel lines) represented in a coordinate form, the respective vectors' coordinates are proportional. Indeed, if there are two collinear vectors, $\vec{a} = (x, y)$ and $\vec{b} = (m, n)$, and $\vec{b} = k \cdot \vec{a}$ for some number k, then clearly, $\dfrac{m}{x} = \dfrac{n}{y} = k$. Finally, to find the length of a vector represented by its coordinates,

we use the formula for the distance between two points. The length of a vector $\overrightarrow{AB}$, where $A(x_A, y_A)$ and $B(x_B, y_B)$, is determined as

$$\left\|\overrightarrow{AB}\right\| = \sqrt{\left(x_B - x_A\right)^2 + \left(y_B - y_A\right)^2}.$$

PROBLEM 7.9.

Prove that for any values of x and y, $10\sqrt{x^2 + y^2} > x + 3y - 2$.

SOLUTION.

Consider two vectors $\vec{a} = (x, y)$ and $\vec{b} = (1, 3)$. The length of each is calculated as

$$\|\vec{a}\| = \sqrt{x^2 + y^2} \text{ and } \|\vec{b}\| = \sqrt{1^2 + 3^2} = \sqrt{10}.$$

The dot product of the vectors $\vec{a}$ and $\vec{b}$ in a coordinate form is

$$\vec{a} \cdot \vec{b} = x \cdot 1 + y \cdot 3 = x + 3y.$$

On the other hand, if these vectors form an angle γ, then their dot product equals

$$\vec{a} \cdot \vec{b} = \|\vec{a}\| \cdot \|\vec{b}\| \cdot \cos \gamma = \sqrt{x^2 + y^2} \cdot \sqrt{10} \cdot \cos \gamma.$$

The range of the function cosine consists of real numbers not exceeding 1 in absolute value. Therefore, no matter what the value of γ is, $|\cos \gamma| \leq 1$. It implies that $\|\vec{a}\| \cdot \|\vec{b}\| \geq \vec{a} \cdot \vec{b}$, which means that $\sqrt{x^2 + y^2} \cdot \sqrt{10} \geq x + 3y > x + 3y - 2$. And we arrive at the sought-after result.

PROBLEM 7.10.

There are given eight non-zero numbers a_1, a_2, a_3, a_4, a_5, a_6, a_7, a_8. Prove that at least one of the following numbers is non-negative:

$$a_1a_3 + a_2a_4,\ a_1a_5 + a_2a_6,\ a_1a_7 + a_2a_8,\ a_3a_5 + a_4a_6,\ a_3a_7 + a_4a_8,\ a_5a_7 + a_6a_8.$$

SOLUTION.

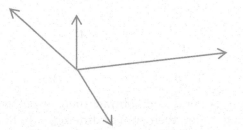

Consider four vectors with coordinates (a_1, a_2), (a_3, a_4), (a_5, a_6), (a_7, a_8) with the common initial point. The six sums in question represent the scalar products of the pairs of these vectors.

The sum of the angles the vectors form is 360°. If at least two of the vectors are perpendicular then their scalar product will be 0 (which satisfies the requirement to be a non-negative number). If no pair of the vectors form a right angle, then there should be at least one pair that the angle they form is an acute angle. The cosine of that angle will be a positive number, and therefore, the respective scalar product of those vectors will be a positive number, which concludes our proof.

One of the isometric transformations of Euclidean plane (distance-preserving transformations) is a *translation* which moves every point of a figure by the same distance in a given direction.

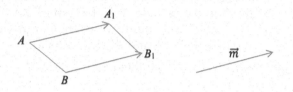

Translation as geometric transformation is associated with a Euclidean vector. Translation preserves the shape and size of any figure, and all respective segments are parallel. In the figure above, vector $\vec{m}$ endows the direction and the magnitude of the specific translation to be applied to segment AB. To obtain images of A and B, the points A_1 and B_1, we move points A and B in the direction of vector $\vec{m}$, so $AA_1 \parallel BB_1 \parallel \vec{m}$, and each of the segments AA_1 and BB_1 has the length of vector $\vec{m}$, $AA_1 = BB_1 = \|\vec{m}\|$.

PROBLEM 7.11.

Prove that the length of a midline of a convex quadrilateral (a segment connecting the midpoints of the opposite sides) is less than or equal to half the sum of the other two sides.

SOLUTION.

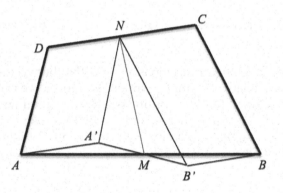

We are given the quadrilateral $ABCD$ in which M and N are the midpoints of AB and DC respectively. We need to prove that $MN \leq \frac{1}{2}(AD + BC)$.

To solve this problem, we will apply translation properties and find the image $A'N$ of AD in translation by vector $\overrightarrow{DN}$ and image $B'N$ of BC in translation by vector $\overrightarrow{CN}$. By definition of translation, $DN = AA'$, $AD = A'N$ and $AD \parallel A'N$, $AA' \parallel DN$, and $CN = BB'$, $BC = B'N$ and $BC \parallel B'N$, $BB' \parallel CN$. If we prove that points A', M, and B' lie on the same straight line, we then will be able to refer to the triangle $A'NB'$ in which two sides are equal to the opposite sides of $ABCD$ in question, $AD = A'N$ and $BC = B'N$, and NM is the median dropped to the third side $A'B'$. Comparing NM to half of the sum of $A'N$ and $B'N$ we will be able to achieve the desired result. So, let's stick to our plan and prove first that A', M, and B' lie on the same straight line (they are collinear points). Consider $\triangle AA'M$ and $\triangle BB'M$.

We see that $AM = MB$ (it is given that M is the midpoint of AB), $AA' = BB'$ ($DN = NC$ because it is given that N is the midpoint of DC, and by translation, $DN = AA'$ and $CN = BB'$), and $\angle MAA' = \angle MBB'$ as interior alternate angles by two parallel lines ($AA' \parallel BB'$ because each is parallel to DC) and transversal AB. Therefore triangles $AA'M$ and $BB'M$ are congruent by Side-Angle-Side. It follows that $\angle AMA' = \angle BMB'$, which implies that these are vertical angles by point M as the common vertex. Therefore, indeed, M lies on $A'B$, i.e. the points A', M, and B' are collinear.

Considering now triangle $A'NB'$, by the parallelogram rule, $\overrightarrow{NM} = \frac{1}{2}\left(\overrightarrow{NA'} + \overrightarrow{NB'}\right)$.

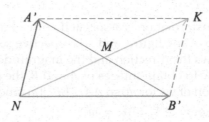

Doing one more translation, this time by vector $\overrightarrow{NB'}$, we get $B'K \parallel NA'$ and $B'K = NA'$. According to the Triangle Inequality Theorem, in triangle $NB'K$,

$$NK < NB' + B'K = NB' + NA'.$$

Therefore, $MN = \frac{1}{2}NK < \frac{1}{2}(NB' + NA')$. Going back to our original quadrilateral $ABCD$ and recalling that $AD = NA'$ and $BC = NB'$, we arrive at the sought-after result that $MN \leq \frac{1}{2}(AD + BC)$.

As we just evidenced, applying translations along with other vectors properties allowed us to get an elegant and relatively simple solution. Moreover, we can make an important observation from our solution that the midline of a quadrilateral will be equal to half the sum of two opposite sides in a quadrilateral only when these sides are parallel.

Also, it worth mentioning that in passing we proved a pretty interesting property that the median of a triangle dropped to the third side is less than half the sum of the other two sides.

PROBLEM 7.12.

Given the diagonals m and n and the altitude h of a trapezoid, find its area.

SOLUTION.

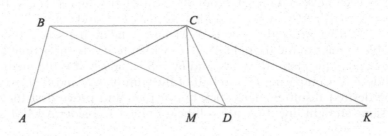

In trapezoid $ABCD$ we are given $AC = m$, $BD = n$, and $CM = h$, where $CM \perp AD$. The goal is to determine the area of the trapezoid. We know that the area of a trapezoid is calculated as half the product of the sum of its bases by its altitude. We are given the length of the altitude and the length of each of the diagonals, but we do not know the lengths of the bases. Should we find the length of each base to apply the formula for the area calculation? Not necessarily. We should not be concerned with each base's length if we manage to find their sum and substitute into the mentioned above formula. This can be done by applying translation properties. We will find the image DK of BC in translation by vector $\overrightarrow{BC}$. We draw a straight line at C parallel to BD till intersection with the extension of AD at K. In the newly formed quadrilateral $BCKD$, $CK \parallel BD$ and $DK \parallel BC$. So, by definition, $BCKD$ is a parallelogram. Therefore, $CK = BD = n$ and $DK = BC$. As we see, in triangle ACK, we know its altitude CM (the altitude of our trapezoid is the altitude of this triangle as well) dropped to AK. We should also note that $AK = AD + DK = AD + BC$. The area of the triangle ACK equals half the product of its base AK by the altitude CM. Therefore, the area of the triangle ACK equals to the area of the trapezoid $ABCD$. So, by finding the area of $\triangle ACK$ we will achieve the desired goal.

Consider the right triangle AMC ($\angle AMC = 90°$). By the Pythagorean Theorem,

$$AM = \sqrt{m^2 - h^2}.$$

Consider the right triangle CMK ($\angle CMK = 90°$). By the Pythagorean Theorem,

$$MK = \sqrt{n^2 - h^2}.$$

Next, $AK = AM + MK = \sqrt{m^2 - h^2} + \sqrt{n^2 - h^2}$.

Finally, the sought-after area is calculated as $S_{ABCD} = S_{ACK} = \dfrac{1}{2} AK \cdot CM = \dfrac{1}{2}\left(\sqrt{m^2 - h^2} + \sqrt{n^2 - h^2}\right) \cdot h$, which is the answer to the problem.

PROBLEM 7.13.

The problem was offered by V. Proizvolov in magazine Квант (in Russian) #2, 1995.

O is a point in the interior of a parallelogram $ABCD$ such that $\angle ABO = \angle ADO$. Prove that $\angle DAO = \angle DCO$.

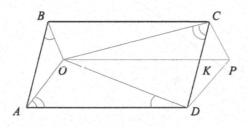

SOLUTION.

Translation by vector $\overrightarrow{BC}$ takes point O into point P such that $OP \parallel BC$ and $PC \parallel OB$. It follows that $OBCP$ is a parallelogram. Then $\angle DCP = \angle ABO$ (as angles between respective parallel lines) and $\angle POD = \angle ADO$ (as alternate interior angles by parallel lines OP and AD and secant OD). Since it is given that $\angle ABO = \angle ADO$, we obtain that $\angle DCP = \angle POD$.

Denote by K the point of intersection of CD and OP. Obviously, $\angle DCP = \angle KCP$. Also, $\angle OKD = \angle PKC$ as vertical angles; therefore, considering triangles OKD and CKP, we conclude that they have two respectively equal angles. This implies that the third respective angles are equal as well, $\angle CPK = \angle ODK$, and consequently, these triangles are similar, that is, $\triangle OKD \sim \triangle CKP$. It follows that $\dfrac{OK}{KD} = \dfrac{CK}{KP}$, which can be rewritten as $OK \cdot KP = CK \cdot KD$. Applying Intersecting Chords Theorem (see the appendix), we arrive at the conclusion that the four points O, C, P, and D are concyclic (four points lie on a common circle), i.e., $OCPD$ is a cyclic quadrilateral. By using this auxiliary circle (not shown in a diagram to not make it too complicated), we can now easily identify congruent inscribed angles subtended by the same chord OD, $\angle DCO = \angle OPD$. Noticing that $\angle OPD = \angle DAO$ as opposite angles in a parallelogram $AOPD$, by transitivity, we arrive at the sought-after result, $\angle DAO = \angle DCO$.

PROBLEM 7.14.

Two squares $ABDE$ and $BCKF$ are built on the sides AB and BC respectively of the given triangle ABC. Prove that the length of the segment DF is twice of the length of the median BP of $\triangle ABC$.

SOLUTION.

Solution is much simplified if we perform translation by vector $\vec{BC}$ of side AB. Such translation takes point A into point L such that $AL \parallel BC$ and $AL = BC$, $BA \parallel CL$ and $BA = CL$. Connecting C and L we get parallelogram $BCLA$.

Since BP is the median of triangle ABC, its extension to point L will form the diagonal of the parallelogram $BCLA$. Considering the angles formed by point B we see that

$$\angle DBF = 360° - 90° - 90° - \angle ABC = 180° - \angle ABC.$$

Also, in parallelogram $ABCL$, $\angle BCL = 180° - \angle ABC$.

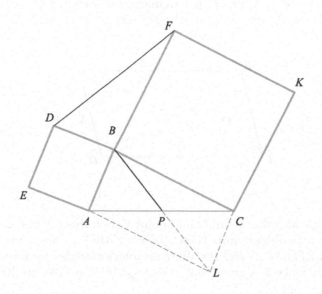

Therefore, $\angle DBF = \angle BCL$, and observing that $BF = BC$ and $DB = BA = CL$, we conclude that triangles DBF and LCB are congruent by Side-Angle-Side. Therefore, $DF = BL$. But we know that $BL = 2BP$, which yields the desired result that $DF = 2BP$.

Translation enabled us to introduce the auxiliary parallelogram. We shifted the focus to comparing the lengths of the segments DF and BL instead of the lengths of the segments DF and BP. This significantly simplified our solution because we immediately obtained the sought-after outcome knowing that the median in question BP is half of the diagonal BL of the auxiliary parallelogram $BCLA$.

Similar trick with auxiliary parallelogram proves useful in the following problem as well.

PROBLEM 7.15.

Prove that the radii of the circumcircles of the triangles AHB, AHC, and BHC, where H is the orthocenter of triangle ABC, are all equal to each other.

SOLUTION.

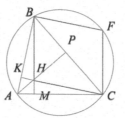

In triangle ABC we draw three altitudes, $AP \perp BC$, $BM \perp AC$, and $CK \perp AB$, and denote H the point of their intersection. Our goal is to prove that the radii of the three circumcircles of the triangles AHB, AHC, and BHC have the same length.

We will apply translation by vector $\overrightarrow{HC}$ of segment BH. The image of BH will be segment FC such that $BH \parallel FC$ and $BH = FC$. Also, as the result of this translation,

$BF \parallel HC$ and $BF = HC$, implying that the newly formed quadrilateral $BFCH$ is a parallelogram. Therefore, its diagonal BC cuts it into two congruent triangles, $\triangle BHC \cong \triangle BFC$. Now, we need to recall that H lies on CK and that $CK \perp AB$. Since $BF \parallel HC$, we obtain that $BF \perp AB$. In a similar way we can prove that $FC \perp AC$. This implies that quadrilateral $ABFC$ is cyclic (the sum of its opposite angles equals 180°) and all four of its vertices A, B, F, and C lie on the same circle. Therefore, we see that triangles ABC and BFC are inscribed in the same circle. As we proved earlier, $\triangle BHC \cong \triangle BFC$. It follows that the radii of circumcircles of triangles ABC and BHC are of the same length. Analogously, we can prove that the radii of circumcircles of triangles AHB and AHC have the same length as the radius of the circumcircle of the triangle ABC and consequently, are equal to each other.

Applying translation in this classic problem allowed getting elegant and straightforward solution without even drawing circumcircles of three triangles in question and locating radius of each such circle on a diagram. Instead, we introduced two auxiliary figures, parallelogram and cyclic quadrilateral, and efficiently utilized their properties to get the desired result. Moreover, we proved an even stronger statement than was required from us in this problem. We have also shown that the radius of the circumcircle of the original triangle ABC equals to the radius of each of the three circumcircles of the triangles formed by connecting its vertices with the orthocenter of the triangle.

How did we come up with the idea for this specific translation? One of the hints is in paying close attention to the most critical of the given conditions. Working with three altitudes in a triangle, one can try finding additional right angles and congruent triangles. Getting a quadrilateral with two opposite right angles pointed out to an idea of a cyclic quadrilateral (sum of the opposite angles of 180° in a quadrilateral identifies it as a cyclic quadrilateral!). In the next step, observing the congruent triangles in an auxiliary parallelogram directed us almost immediately to the sought-after result.

In conclusion of this chapter we need to stress that we concentrated on practical applications of Euclidean vectors in problem-solving in Euclidean space. The definition of a Euclidean vector should not be confused with the more general definition of a vector in pure mathematics as any element of a vector space, which is a more advanced topic and is beyond the scope of our book.

8

Cartesian Coordinates in Problem Solving

With me, everything turns into mathematics.

René Descartes

The prominent French mathematician and philosopher René Descartes (1596–1650) developed and introduced the Cartesian coordinate system, which allows to uniquely define a point in a plane or three-dimensional space through its coordinates, two or three numbers respectively. Descartes's ingenious discovery provided a method for expressing geometric shapes by algebraic equations. Such equations are satisfied by the coordinates of any point lying on a geometric shape described by an equation. For instance, an equation of a straight line in a plane is $y = ax + b$ or an equation of a circle with the center at point O with the coordinates a and b, i.e. $O(a,b)$, and radius r is $(x - a)^2 + (y - b)^2 = r^2$. On one hand, Cartesian coordinates allow us to solve geometric problems using algebra, and on the other hand, to visualize many algebraic relationships. They also provide geometrical interpretations for such branches of mathematics as complex analysis, multivariable calculus, group theory, and many more.

Descartes's work establishing analytic geometry (it is also referred to as Cartesian geometry) greatly influenced the development of Newton's and Leibniz's discoveries of calculus. The two-dimensional Cartesian plane was later generalized into the concept of vector spaces. Cartesian coordinates have wide recognition and application in many areas of science, from astronomy and physics to engineering, and also in computer graphics, computer-aided geometric design, and other geometry-related data processing.

In this chapter we will investigate several examples illustrating that Cartesian coordinates are essential tools in establishing connecting links between algebra and geometry.

Using the Cartesian coordinate system technique we can find elegant and relatively simple solutions to many difficult problems, even those offered on mathematical Olympiads. In certain cases, it gives an enlightening new glance at a problem or even paves a straightforward path to the solution. Moreover, even though such solutions are unusual and not typically anticipated, they are simple to understand.

In our following discussion, we expect readers to be fairly familiar with the basic definitions and properties of the orthogonal Cartesian coordinate system in a two and three-dimensional space, where the coordinates can be found as the numeral positions of the perpendicular projections of the point onto the coordinate axes.

We will start with a few basic problems showing the application of the Cartesian coordinate system in finding graphical solutions to the systems of equations and inequalities.

DOI: 10.1201/9781003359500-8

PROBLEM 8.1.

Solve the system of equations:

$$\begin{cases} x^2 - 2x + y^2 - 3 = 0, \\ -\dfrac{3}{2}x - y = -6. \end{cases}$$

<u>SOLUTION.</u>

This system is easily solvable using the algebraic substitution technique. Express y through x from the second equation and then substitute this expression into the first equation. Solve a quadratic equation in x. Substitute your answers into the second equation and solve for y. We suggest that the readers complete these steps independently.

Here we will demonstrate the geometrical solution through graphs. The solutions of such a system are all the ordered pairs that satisfy both equations. To solve a system of equations graphically, we graph both equations in the same coordinate system. The solutions to the system will be in the points where the two graphs intersect.

To start, we will modify the first equation by completing the square, and will express y in terms of x in the second equation.

$$x^2 - 2x + y^2 - 3 = x^2 - 2x + 1 + y^2 - 4 = (x-1)^2 + y^2 - 4.$$

Now, we can rewrite the system as

$$\begin{cases} (x-1)^2 + y^2 = 2^2, \\ y = -\dfrac{3}{2}x + 6. \end{cases}$$

Graph of the first equation is the circle with center at $M(1, 0)$ and radius 2. Graph of the second equation is the straight line with the slope $-\dfrac{3}{2}$, and 6 is the intercept of line on the Y-axis.

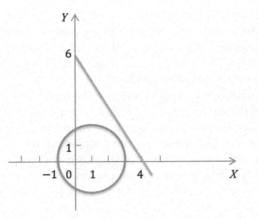

As we can see, there are no points of intersection of the graphs. Therefore, the system of equations has no solutions. Generally, if a straight line is tangent to a circle, then the system has one solution, the coordinates of the point of tangency. In case when a straight line intersects a circle in two points, there are two solutions, the coordinates of each of the common points of intersection. One of the drawbacks of using graphical method for solving systems of equations is the inability sometimes to determine the precise coordinates of the common points of intersection. So, verification of your findings by substituting the coordinates of the points of intersection into each equation of the system is required.

PROBLEM 8.2.

Solve the system of inequalities:

$$\begin{cases} y \le \dfrac{1}{2}x, \\ x \ge 0. \end{cases}$$

SOLUTION.

Consider linear function $y = \dfrac{1}{2}x$ and sketch its graph below.

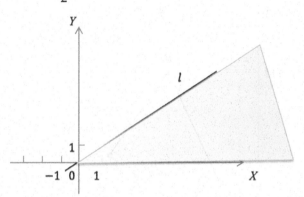

Solutions of our system are all the points located below straight line l (its equation is $y = \dfrac{1}{2}x$) and above X-axis, that is, the green area in a diagram above.

Before we proceed further, let's recall that having the coordinates of the points $A(x_A, y_A)$ and $B(x_B, y_B)$ in a two-dimensional space, the distance between them is calculated by the formula $AB = \sqrt{(x_B - x_A)^2 + (y_B - y_A)^2}$.

Each point in a three-dimensional space has three coordinates, hence, the length of the segment AB, where $A(x_A, y_A, z_A)$ and $B(x_B, y_B, z_B)$, is calculated as $AB = \sqrt{(x_B - x_A)^2 + (y_B - y_A)^2 + (z_B - z_A)^2}$.

The distance from a point $A(x_1, y_1, z_1)$ to a plane in a three-dimensional space defined by the equation $mx + ny + qz + p = 0$ is calculated by the formula

$$\dfrac{|mx_1 + ny_1 + qz_1 + p|}{\sqrt{m^2 + n^2 + q^2}}.$$

PROBLEM 8.3.

Find the minimum value of the expression without resorting to calculus means

$$\sqrt{(x-5)^2 + (y-12)^2} + \sqrt{x^2 + y^2}.$$

SOLUTION.

Don't be bogged down with confusing expression that is given. We could fairly easily resolve this problem by examining it from a non-standard point of view. Although the solution is not what would occur to many readers at first thought, the problem can be readily solved using only the tools of the Cartesian coordinates technique.

Let's consider the function $f(x,y) = \sqrt{(x-5)^2 + (y-12)^2} + \sqrt{x^2 + y^2}$.

Translating the sum on the right-hand side in terms of the Cartesian coordinates, we will obtain the sum of two distances from some point $A(x,y)$ to point $B(5, 12)$ and to origin $O(0, 0)$.

Indeed, the distance from A to B is calculated in a coordinates format as $AB = \sqrt{(x-5)^2 + (y-12)^2}$, and the distance from A to O is $AO = \sqrt{x^2 + y^2}$.

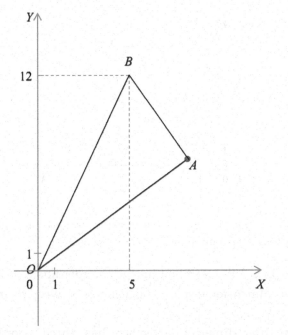

Applying the Triangle inequality theorem that states that the sum of any two sides in a triangle must be greater than the length of the third side, we have that $AB + AO \geq BO$.

Using the coordinates of B and O, we calculate the distance BO as $BO = \sqrt{5^2 + 12^2} = \sqrt{169} = 13$. Therefore, the minimum of $f(x,y)$ is 13 and this is the value we seek.

Using similar reasoning, we can now easily tackle Problem 8.4 offered on the XIV International Olympiad "Intellectual Marathon" and discussed in magazine Квант, #3, 2007 (in Russian).

PROBLEM 8.4.

Find the minimum value of the function

$$f(x) = \sqrt{x^2 - 6x + 13} + \sqrt{x^2 - 14x + 58} .$$

SOLUTION.

Completing the square under each radical sign, we can rewrite this function as

$$f(x) = \sqrt{(x-3)^2 + 4} + \sqrt{(x-7)^2 + 9} = \sqrt{(x-3)^2 + (\pm 2)^2} + \sqrt{(x-7)^2 + (\pm 3)^2}.$$

Now, referring to the Cartesian coordinates, we can see that the right-hand side of the last expression is the sum of the distances from point $A(x,0)$ to points $B_1(3, -2)$ and $C_1(7, 3)$ or to points $B_2(3, 2)$ and $C_2(7, -3)$. Point $A(x,0)$ lies on X-axis (because its ordinate is 0). Therefore, applying the Triangle inequality theorem, we conclude that the minimum of the sums AB and AC will be attained when A lies on the segment BC. But this can be possible only when B and C lie in the different quadrants of the plane, the first and the fourth. That's why we consider only two scenarios when ordinates of B and C have opposite signs.

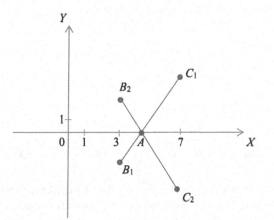

Calculating B_1C_1, we get $B_1C_1 = \sqrt{(7-3)^2 + (3-(-2))^2} = \sqrt{41}$.

Calculating B_2C_2, we get $B_2C_2 = \sqrt{(7-3)^2 + (-3-2)^2} = \sqrt{41}$.

So, in either case we get the same outcome of $\sqrt{41}$. This is the sought-after minimum value of the function $f(x)$.

PROBLEM 8.5

In the equation $y = \sqrt{2x - x^2 + 15}$ find the range of possible values of y.

SOLUTION.

First, we will determine the domain of the function $y = \sqrt{2x - x^2 + 15}$.

It consists of all real numbers such that $-x^2 + 2x + 15 \geq 0$ or, equivalently,

$$x^2 - 2x - 15 \leq 0.$$

Finding the roots of the quadratic trinomial, we can factor it as $(x+3)(x-5) \le 0$.

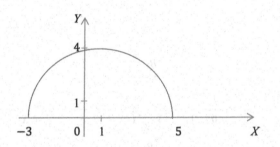

The domain of the function is $x \epsilon [-3, 5]$.

Observing that the square root of a number is always non-negative, we obtain that our equation is equivalent to the system

$$\begin{cases} y^2 = 2x - x^2 + 15, \\ -3 \le x \le 5, \\ y \ge 0. \end{cases}$$

Completing the square on the right-hand side, we can rewrite the first equation as $(x-1)^2 + y^2 = 4^2$. In a Cartesian coordinate system, this is the equation of the circle with center at $O(1, 0)$ and radius 4. Now, we can illustrate our system in the figure below, considering x such that $-3 \le x \le 5$ and y such that $y \ge 0$.

Our goal is to find the intersection of the three sets of numbers in our system. The points with coordinates that obey these conditions constitute the semicircle above X-axis with the ordinates from 0 to 4. In other words, the range of y satisfying the given equation is the set of numbers from 0 to 4, $y \epsilon [0, 4]$, which is the answer to the problem.

There are many geometry problems in which introducing the Cartesian coordinates facilitates the solution or opens a new unexpected and efficient approach.

PROBLEM 8.6.

Prove that for any randomly selected point M interior of a rectangle $ABCD$ the following equality holds

$$MA^2 + MC^2 = MB^2 + MD^2.$$

SOLUTION.

We will introduce a Cartesian coordinate system in such a way that A is at origin, AD lies on X-axis, and AB lies on Y-axis. Denote $AD = a$ and $AB = b$. Then the initial positions of each vertex of our rectangle get the coordinates, $A(0,0)$, $B(0,b)$, $D(a, 0)$, and $C(a, b)$. For randomly selected point M interior of $ABCD$, we will denote its coordinates

m and n, $M(m, n)$. The next step is to calculate all the distances in question through the coordinates of the points.

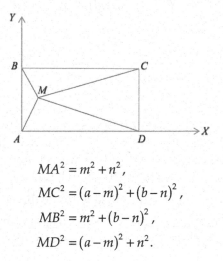

$$MA^2 = m^2 + n^2,$$
$$MC^2 = (a-m)^2 + (b-n)^2,$$
$$MB^2 = m^2 + (b-n)^2,$$
$$MD^2 = (a-m)^2 + n^2.$$

Hence,

$$MA^2 + MC^2 = m^2 + n^2 + (a-m)^2 + (b-n)^2 \text{ and } MB^2 + MD^2 = m^2 + (b-n)^2 + (a-m)^2 + n^2.$$

The right-hand sides of each of these sums are identical.
   Therefore, indeed, regardless of where M is interior of $ABCD$, the equality holds:

$$MA^2 + MC^2 = MB^2 + MD^2.$$

One final remark – applying similar approach, one can easily extend this problem and prove that the same equality holds for any location of point M, regardless it is inside or outside of $ABCD$.

PROBLEM 8.7.

Prove that the sum of the squares of the distances from a point M on the diameter of a circle to the endpoints of any chord parallel to this diameter is a constant number.

SOLUTION.

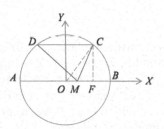

Consider a circle with center O and diameter AB. Point M lies on AB. Draw a chord $CD \parallel AB$, and connect M with C and D. Our goal is to prove that $MC^2 + MD^2$ is some constant number regardless of the location of CD parallel to AB. Let's introduce a coordinate system on the plane such that the center of the circle O coincides with origin, AB lies on X-axis, and

Y-axis is perpendicular to *AB* (passing, obviously, through *O*). The initial positions of the points get the coordinates $M(m, 0)$, $C(a, b)$ and $D(-a, b)$.

Then $MC^2 = (a-m)^2 + b^2$ and $MD^2 = (-a-m)^2 + b^2 = (a+m)^2 + b^2$. Adding these two equalities gives

$$MC^2 + MD^2 = (a-m)^2 + b^2 + (a+m)^2 + b^2 = a^2 - 2am + m^2 + 2b^2 + a^2 + 2am + m^2 =$$
$$2a^2 + 2m^2 + 2b^2 = 2(a^2 + b^2) + 2m^2 \ (*)$$

It's not hard to see that $a^2 + b^2 = r^2$, where r represents the radius of the circle. Indeed, if we drop *CF* perpendicular to *AB*, then in the right triangle *CFO* we have $CF = b$ and $OF = a$. By the Pythagorean Theorem, $CO^2 = CF^2 + OF^2$. Noticing that *CO* is the radius, i.e. $CO = r$, and substituting values for *CF* and *OF* we get that $a^2 + b^2 = r^2$. We can now go back to (*) and modify it as $MC^2 + MD^2 = 2(r^2 + m^2)$. Since r and m are some predetermined fixed numbers (r is the radius of the circle with the diameter *AB*, i.e. $r = \frac{1}{2} AB$, and m is the distance form *M* to *O*), then $MC^2 + MD^2$ is a constant number regardless of how we draw a chord parallel to *AB*, which is what we wished to prove.

PROBLEM 8.8.

A straight line that goes through random point *M* laying inside a circle whose center is *O* and radius *r* intersects this circle at points *A* and *B*.

Prove that $MA \cdot MB = |OM^2 - r^2|$.

SOLUTION.

The proof of the assertion of the problem using conventional techniques is far from easy. We will examine a relatively simple and elegant proof applying the coordinate system technique. The beauty and instructional benefit of this approach are manifested in the simplicity of the algebraic steps used to prove the geometric property in question.

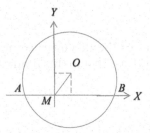

We will plot points *A* and *B* on a coordinate plane with origin at point *M* such that *AB* lies on *X*-axis. Introducing the coordinates, we have $M(0, 0)$ and $O(a, b)$ (*a* and *b* are some real numbers). Each of the points *A* and *B* lies on the circle, thus the coordinates of each point satisfy the equation of the circle $(x - a)^2 + (y - b)^2 = r^2$. On the other hand, the same coordinates satisfy the equation of *X*-axis, which is $y = 0$. So, to find the abscissas of *A* and *B*, we need to solve the following system of equations:

$$\begin{cases} (x-a)^2 + (y-b)^2 = r^2, \\ y = 0. \end{cases}$$

Substituting $y = 0$ into the first equation gives $x^2 - 2ax + (a^2 + b^2) - r^2 = 0$.

The last equation is quadratic in x. Noticing that $MO = \sqrt{a^2 + b^2}$ (the distance between M and O expressed through their coordinates), we can rewrite the equation as

$$x^2 - 2ax + (MO^2 - r^2) = 0.$$

The roots of this quadratic equation represent the abscissas of A and B, the points of intersection of the circle with X-axis. Geometrically, the $|x_1|$ is the distance from A to M on X-axis and $|x_2|$ is the distance from B to M on X-axis, or in other words,

$$|x_1 \cdot x_2| = MA \cdot MB.$$

All we have to do now is to refer to Vieta's formulas applied to quadratic equation and observe that $x_1 \cdot x_2 = MO^2 - r^2$. It implies that $MA \cdot MB = |OM^2 - r^2|$, which is the relationship we set to develop.

We should point out that in fact, we've managed to get by without any referral to a diagram. Here we confined ourselves to working with coordinates only, which was sufficient to demonstrate our method of solving another purely geometrical problem by applying algebraic techniques.

PROBLEM 8.9.

In a tetrahedron $DABC$ there are three right angles by the vertex D, and $DA = a$, $DB = b$, $DC = c$. Find the altitude of $DABC$.

SOLUTION.

Contrary to the previous problem, the hint to the solution of this problem is hidden in a diagram depicting the problem's conditions. Moreover, it helps to judiciously introduce a coordinate system on the plane, so the initial positions of the points get the most beneficial coordinates for subsequent calculations.

Sometimes to solve a geometry problem you have merely to stare at its diagram until all of a sudden, you grasp the new idea, and the solution pops up at you in all its clarity. How about if we turn around the tetrahedron in Figure 8.1 (where it is shown in standard fashion), and instead consider the same tetrahedron in Figure 8.2. This will be a hint to introduce a Cartesian coordinate system in a three-dimensional space with origin at point D and X-axis aligned along edge DA, Y-axis along edge DB, and Z-axis along edge DC.

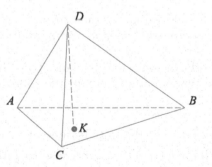

FIGURE 8.1

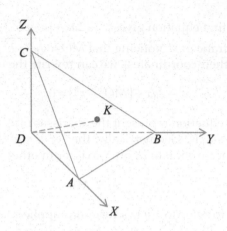

FIGURE 8.2

Knowing the length of each of the edges DA, DB, and DC, we can assign the respective coordinates to the vertices of our tetrahedron, $D(0, 0, 0)$, $A(a, 0, 0)$, $B(0, b, 0)$ and $C(0, 0, c)$. Our problem can be transformed now to a problem about finding the distance from a point to a plane; in other words, we need to determine the distance DK from point D to plane (ABC).

In three-dimensional space the general equation of a plane is written as

$$mx + ny + qz + p = 0.$$

The distance from a random point $M(x_1, y_1, z_1)$ to a plane defined by the above equation is calculated by the formula

$$\frac{\left|mx_1 + ny_1 + qz_1 + p\right|}{\sqrt{m^2 + n^2 + q^2}}.$$

Since we will be looking for the distance from $D(0, 0, 0)$ to the plane ABC, the above formula is modified to

$$\frac{\left|m \cdot 0 + n \cdot 0 + q \cdot 0 + p\right|}{\sqrt{m^2 + n^2 + q^2}} = \frac{|p|}{\sqrt{m^2 + n^2 + q^2}} \quad (*)$$

In fact, we don't even need to determine the equation of the plane ABC. For our purposes it suffices to calculate the coefficients m, n, q, and p, and then substitute their values into (*).

Each of the points A, B, and C belongs to this plane. Thus the coordinates of each must satisfy the equation of the plane, and we obtain the following system of three linear equations:

$$\begin{cases} m \cdot a + n \cdot 0 + q \cdot 0 + p = 0, \text{ because } A \in (ABC) \\ m \cdot 0 + n \cdot b + q \cdot 0 + p = 0, \text{ because } B \in (ABC) \\ m \cdot 0 + n \cdot 0 + q \cdot c + p = 0, \text{ because } C \in (ABC). \end{cases}$$

This system is simplified to

$$
\begin{cases}
m = -\dfrac{p}{a}, \\[2mm]
n = -\dfrac{p}{b}, \\[2mm]
q = -\dfrac{p}{c}.
\end{cases}
$$

Therefore, $DK = \dfrac{|p|}{\sqrt{m^2 + n^2 + q^2}} = \dfrac{|p|}{\sqrt{\dfrac{p^2}{a^2} + \dfrac{p^2}{b^2} + \dfrac{p^2}{c^2}}} = \dfrac{1}{\sqrt{\dfrac{1}{a^2} + \dfrac{1}{b^2} + \dfrac{1}{c^2}}}.$

Denoting the altitude by h, we can rewrite the formula in a different way, easy to remember for the general case:

$$
\frac{1}{h^2} = \frac{1}{a^2} + \frac{1}{b^2} + \frac{1}{c^2}.
$$

The Cartesian coordinates' introduction we've used in this problem was very visual and intuitively clear. The previous problems lack this easy touch. The question arises about how to decide what is the most beneficial and effective way to plot your figure on a coordinate plane or in a three-dimensional space. It all depends on the conditions of a problem one is solving. With a small stretch of the imagination one can try different configurations, finally evolving with the best one leading to the easiest calculations, as will be demonstrated in the following problems 8.10 and 8.11.

PROBLEM 8.10.

There are given two flat angles of a trihedral angle with vertex O, $\angle AOB = \alpha$ and $\angle AOC = \beta$ ($\alpha < 90°$ and $\beta < 90°$). Find the third flat angle if it is known that the opposite of this flat angle dihedral angle is a right angle.

SOLUTION.

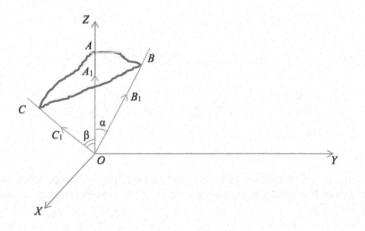

We have already evidenced in the previous chapters that a problem about finding an angle often is simplified to finding a trigonometric function of an angle in question. Studying vectors in the previous chapter we frequently applied the dot product of two vectors when we wanted to find the angle between them. Using these ideas and applying vector algebra techniques and introducing the Cartesian coordinates make this problem relatively simple.

For two vectors $\vec{a}(x_1, y_1, z_1)$ and $\vec{b}(x_2, y_2, z_2)$ in a three-dimensional space, their dot product is defined as $\vec{a} \cdot \vec{b} = \|\vec{a}\| \cdot \|\vec{b}\| \cdot \cos\gamma = x_1 x_2 + y_1 y_2 + z_1 z_2$, from which $\cos\gamma = \dfrac{x_1 x_2 + y_1 y_2 + z_1 z_2}{\|\vec{a}\| \cdot \|\vec{b}\|}$, or substituting $\|\vec{a}\| \cdot \|\vec{b}\|$ expressed in a coordinate form, we get

$$\cos\gamma = \frac{x_1 x_2 + y_1 y_2 + z_1 z_2}{\sqrt{x_1^2 + y_1^2 + z_1^2} \cdot \sqrt{x_2^2 + y_2^2 + z_2^2}}.$$

There are two supplementary angles formed when two straight lines intersect. Since we always choose $0 \le \gamma \le \dfrac{\pi}{2}$, we can rewrite the last expression as

$$\cos\gamma = \frac{\left| x_1 x_2 + y_1 y_2 + z_1 z_2 \right|}{\sqrt{x_1^2 + y_1^2 + z_1^2} \cdot \sqrt{x_2^2 + y_2^2 + z_2^2}}.$$

All we need to do now is to decide how to introduce the Cartesian coordinates so we can avoid tedious calculations and get an easy result.

We will place the given trihedral angle in the Cartesian coordinate system in such a way that vertex O of the trihedral angle coincides with origin, Z-axis goes along the edge OA of the right dihedral angle, X-axis lies in plane AOC and Y-axis lies in plane AOB. The goal is to find the measure of the angle BOC.

Let's introduce three unit vectors (the length of each equals to 1) $\overrightarrow{OA_1}$, $\overrightarrow{OB_1}$, and $\overrightarrow{OC_1}$ lying respectively on the edges OA, OB, and OC. Finding their coordinates, we see that

$$\overrightarrow{OA_1} = (0,0,1), \ \overrightarrow{OB_1} = (0, \sin\alpha, \cos\alpha), \text{ and } \overrightarrow{OC_1} = (\sin\beta, 0, \cos\beta).$$

Finding the dot product of the vectors $\overrightarrow{OB_1}$ and $\overrightarrow{OC_1}$ and expressing the cosine of the angle between them yields

$$\cos\angle BOC = \frac{\overrightarrow{OB_1} \cdot \overrightarrow{OC_1}}{\left|\overrightarrow{OB_1}\right| \cdot \left|\overrightarrow{OC_1}\right|} = \frac{0 \cdot \sin\beta + \sin\alpha \cdot 0 + \cos\alpha \cdot \cos\beta}{1 \cdot 1} = \cos\alpha \cdot \cos\beta.$$

Hence, $\angle BOC = \arccos(\cos\alpha \cdot \cos\beta)$, which concludes our solution.

PROBLEM 8.11.

Given that in a regular triangular prism the length of its side edge (they are all equal) is equal to the length of the base's side, find the angle between the nonintersecting diagonals of the two side faces.

SOLUTION.

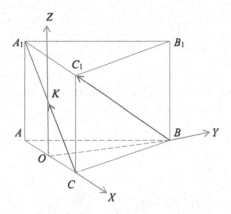

Recall that a regular triangular prism is a triangular prism whose bases are regular triangles and the side faces are rectangles. It is given that $AA_1 = BB_1 = CC_1 = AB = AC = BC$. The goal is to find the measure of the angle between CA_1 and BC_1, the diagonals lying on two skew lines in the faces AA_1C_1C and BB_1C_1C. A three-dimensional problem about finding an angle between two skew lines is not a simple problem to solve. This problem has several different purely geometrical solutions, which we leave to the readers to explore. We will illustrate another, quite simple and elegant solution that seems more attractive in our view.

Perhaps the best will be to plot the mid-point O of AC as origin, and align X-axis along AC, Y-axis along OB, and Z-axis along OK, where K is the mid-point of A_1C and $OK \parallel AA_1$. We then can find $\cos \gamma$ for the angle γ between vectors $\overrightarrow{CK}$ and $\overrightarrow{BC_1}$ (clearly, it is the same angle as between vectors $\overrightarrow{CA_1}$ and $\overrightarrow{BC_1}$) with relatively simple calculations.

Designating the length of each edge (remember, it is given that all the edges are of the same length) 1, we see that $OC = \dfrac{1}{2}$ and we can easily find that $OB = \dfrac{\sqrt{3}}{2}$ (applying the Pythagorean Theorem to triangle COB). Therefore, we can now set our points with their coordinates as $O(0, 0, 0)$, $K\left(0, 0, \dfrac{1}{2}\right)$, $C\left(\dfrac{1}{2}, 0, 0\right)$, $B\left(0, \dfrac{\sqrt{3}}{2}, 0\right)$, and $C_1\left(\dfrac{1}{2}, 0, 1\right)$. It follows that $\overrightarrow{CK} = \left(-\dfrac{1}{2}, 0, \dfrac{1}{2}\right)$ and $\overrightarrow{BC_1} = \left(\dfrac{1}{2}, -\dfrac{\sqrt{3}}{2}, 1\right)$ and respectively,

$$\cos \gamma = \frac{\left| -\dfrac{1}{2} \cdot \dfrac{1}{2} + 0 \cdot \left(-\dfrac{\sqrt{3}}{2}\right) + \dfrac{1}{2} \cdot 1 \right|}{\sqrt{\dfrac{1}{4} + 0 + \dfrac{1}{4}} \cdot \sqrt{\dfrac{1}{4} + \dfrac{3}{4} + 1}} = \frac{\dfrac{1}{4}}{1} = \frac{1}{4}.$$

Thus, we conclude that the angle between the nonintersecting diagonals of the two side faces is $\arccos \dfrac{1}{4}$, the value we were looking for.

Combinatorics is considered by many to be the most difficult subject in mathematics. Some attribute this impression to the fact that it deals with discrete phenomena as opposed to continuous phenomena. The argument is made that you don't "feel" a combinatorial problem, or, in other words, you can't visualize it. In certain cases, the Cartesian coordinate system lightens up the whole picture and paves the way to unexpected instructive, and elegant solutions.

PROBLEM 8.12.

A die is rolled twice. What is the probability of getting a number not less than 5 at least once?

SOLUTION.

It is natural to assume that each of the six faces has the same chance of turning up. So, the probability of rolling a 5 is $\frac{1}{6}$; the probability of rolling a number not less than 5 is $\frac{2}{6} = \frac{1}{3}$. Bear in mind that we pick only the favorable outcomes out of all possible outcomes.

Since we are rolling a die twice, let's consider those sets – favorable outcomes as consisting of pairs (x, y), where x is the number that turns up in the first roll and y in the second.

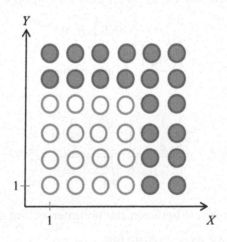

Note that $1 \le x \le 6$ and $1 \le y \le 6$. There are $6 \cdot 6 = 36$ such pairs in total. How about to represent these pairs as a 6×6 circle array located on a coordinate plane (the unit cell with coordinates x and y represents the pair (x, y))?

The favorable outcomes (corresponding pairs) where $x \geq 5$ and $y \geq 5$ are colored blue in the figure above. We have 20 out of total of 36, so the probability of getting a number not less than 5 is $\dfrac{20}{36} = \dfrac{5}{9}$, which is the number we seek.

To conclude our explorations, we will consider a very interesting and enthralling problem that was offered by A. Kanel-Belov (А. Канель-Белов in Russian) in the Russian magazine Квант #4, 2013, problem M2311.

PROBLEM 8.13.

A fly is flying inside the unitary cube (a cube with edges of length 1). What is the minimum length of her flying trajectory visiting all the faces and coming back to the initial starting point?

SOLUTION.

The problem at first looks daunting.

To investigate the trajectory of a flight, we need to consider a closed broken line (the fly travels back to her initial point). The sum of the lengths of all the segments of such a broken line will be the length of her flight.

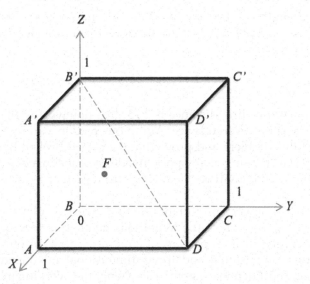

To simplify our calculations, we will introduce the Cartesian coordinates in a three-dimensional space, with point B at origin, and the axes X, Y, and Z aligned at the sides of our cube BA, BC, and BB' respectively. Then the vertices of the cube can be presented with their coordinates as follows: $A(1,0,0)$, $B(0,0,0)$, $C(0,1,0)$, and $B'(0,0,1)$. Let our broken line consists of n segments. Let's denote (x_k, y_k, z_k), $k = 1,2,\ldots, n$ the coordinates of the vectors defining the segments of this broken line. Then the length of the k-s segment is $d_k = \sqrt{x_k^2 + y_k^2 + z_k^2}$, and the length of the broken line will be

$$D = d_1 + d_2 + \cdots + d_n.$$

Let $X = |x_1| + \cdots + |x_n|$ (this is the sum of projections of the fly on X-axis), and similarly, $Y = |y_1| + \cdots + |y_n|$ and $Z = |z_1| + \cdots + |z_n|$.

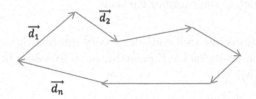

Consider the initial location of the fly at a point F in the plane $(B'BC)$. The equation of this plane is $x = 0$. The shortest distance from F to the opposite plane expressed by the equation $x = 1$ (remember, fly is traveling inside of unitary cube) will be a perpendicular dropped from F to that plane. Eventually fly will return to the position at F. Since she is going to visit every face of the cube, clearly, her total covered distance will be greater or equal than her traveling distance if she would just get back from plane $x = 1$ to F in plane $x = 0$. This minimal distance back and force is the distance between the opposite faces of the unitary cube and it equals 2. Therefore, $X \geq 2$. Similarly, $Y \geq 2$ and $Z \geq 2$. In case when a fly is not located initially in any faces of the cube, she will need to land on one of the faces anyway at her first stop, so the total covered distance will be even greater than we considered above.

The sum of all the vectors with coordinates $(|x_k|, |y_k|, |z_k|)$ equals a vector with coordinates (X, Y, Z). The length of a broken line is greater than or equal the distance between its ends. Therefore, the sum of the lengths of these vectors is not less than the length of the resulting vector-sum, hence

$$d_1 + d_2 + \cdots + d_n \geq \sqrt{X^2 + Y^2 + Z^2} \geq \sqrt{2^2 + 2^2 + 2^2} = \sqrt{12} = 2\sqrt{3}.$$

Therefore, the minimal traveling distance is $2\sqrt{3}$. It is noteworthy that this is the length of the diagonal of the cube covered twice. Ambitious readers can investigate a pure geometrical solution of this problem and prove that the closed broken line would have such a length if a fly will start moving at any point inside the cube in a direction parallel to the diagonal of the cube and then will be "bouncing" from each of its faces.

In Chapter 3 we touched on very important role that inversion plays in the Poincaré Disk Model and discussed the notion of the hyperbolic geometry. Speaking about Cartesian coordinates, in conclusion, it is important to emphasize that all our definitions in this chapter relate to two-dimensional and three-dimensional Euclidean space of classical geometry rather than non-Euclidean geometries. Generally, working in real n-space R^n, the Cartesian coordinates of a point represent a finite order of real numbers (so-called n-tuple) associated with that point and they allow us to locate that point in the Euclidean space.

9

Inequalities Wonderland

Mathematics is the tool specially suited for dealing with abstract concepts of any kind and there is no limit to its power in this field.

Paul Dirac

Inequalities are very important and are broadly used in all branches of mathematics. Learning about them paves the way to a deeper understanding of various topics in algebra, calculus, geometry, and number theory.

Solving problems concerning inequalities involve immensely different techniques than solving equations. This is like operating in a different universe. In most cases, you are not required to find a specific value, but rather the range of values satisfying an inequality. To prove inequality holds true, one needs to determine a technique allowing one to verify the validity of a statement in question satisfying restrictions imposed on variables. We will go over several such techniques and reveal useful tricks in dealing with inequalities proofs. In this chapter, we will also show how inequalities can be applied (sometimes, unexpectedly) to solving many interesting problems that originally have nothing to do with inequalities. We will demonstrate how crucial it is to properly utilize them in revealing valuable connections between math disciplines and taking advantage of those connections for making solutions to complicated problems not just manageable, but efficient and elegant.

One of the well-known classic inequalities that prove useful whenever we try to compare the numerical expressions or assess their upper or lower bounds is *AM-GM Inequality* introduced and frequently referred to in previous chapters:

The arithmetic mean of any n nonnegative real numbers is greater than or equal to their geometric mean. The two means are equal if and only if all the numbers are equal:

$$\frac{a_1 + a_2 + \cdots + a_n}{n} \geq \sqrt[n]{a_1 \cdot a_2 \cdot \ldots \cdot a_n}.$$

PROBLEM 9.1.

Which one is the greater of two numbers $a = \sqrt[3]{25} + \sqrt[3]{9}$ or $b = \sqrt[6]{14350}$?

SOLUTION.

Applying AM-GM Inequality to evaluate a, gives $a = \sqrt[3]{25} + \sqrt[3]{9} \geq 2 \cdot \sqrt[6]{25 \cdot 9} = \sqrt[6]{64 \cdot 25 \cdot 9} = \sqrt[6]{14400} > \sqrt[6]{14350}$. Therefore, we conclude that $a > b$.

DOI: 10.1201/9781003359500-9

PROBLEM 9.2.

Given that all the numbers $x_1, x_2, \ldots, x_k$ are of the same sign (either all positive or all negative), prove that $\dfrac{x_1}{x_2} + \dfrac{x_2}{x_3} + \dfrac{x_3}{x_4} + \cdots + \dfrac{x_{k-1}}{x_k} + \dfrac{x_k}{x_1} \geq k.$

SOLUTION.

Clearly, each fraction on the left-hand side of the expression in question is a positive number because the nominator and denominator of each are of the same sign. Therefore, we can apply AM-GM Inequality and obtain that

$$\frac{x_1}{x_2} + \frac{x_2}{x_3} + \frac{x_3}{x_4} + \cdots + \frac{x_{k-1}}{x_k} + \frac{x_k}{x_1} \geq k \cdot \sqrt[k]{\frac{x_1}{x_2} \cdot \frac{x_2}{x_3} \cdot \frac{x_3}{x_4} \cdots \frac{x_{k-1}}{x_k} \cdot \frac{x_k}{x_1}} = k.$$

PROBLEM 9.3.

Prove that for positive numbers a, b, and c, $\dfrac{bc}{a} + \dfrac{ac}{b} + \dfrac{ab}{c} \geq a + b + c.$

SOLUTION.

Applying AM-GM Inequality three times we have that

$$\frac{1}{2}\left(\frac{bc}{a} + \frac{ac}{b}\right) \geq \sqrt{\frac{bc}{a} \cdot \frac{ac}{b}} = \sqrt{c^2} = |c| = c \text{ (because } c > 0),$$

$$\frac{1}{2}\left(\frac{bc}{a} + \frac{ab}{c}\right) \geq \sqrt{\frac{bc}{a} \cdot \frac{ab}{c}} = \sqrt{b^2} = |b| = b \text{ (because } b > 0),$$

$$\frac{1}{2}\left(\frac{ac}{b} + \frac{ab}{c}\right) \geq \sqrt{\frac{ac}{b} \cdot \frac{ab}{c}} = \sqrt{a^2} = |a| = a \text{ (because } a > 0).$$

Adding the above inequalities gives

$$\frac{1}{2}\left(\frac{bc}{a} + \frac{ac}{b}\right) + \frac{1}{2}\left(\frac{bc}{a} + \frac{ab}{c}\right) + \frac{1}{2}\left(\frac{ac}{b} + \frac{ab}{c}\right) \geq c + b + a.$$

Simplifying the expression on the left-hand side leads to the sought-after result of

$$\frac{bc}{a} + \frac{ac}{b} + \frac{ab}{c} \geq a + b + c.$$

PROBLEM 9.4.

Prove that

$$\left(\frac{1}{\sqrt{1 \cdot 1978}} + \frac{1}{\sqrt{2 \cdot 1977}} + \cdots + \frac{1}{\sqrt{n \cdot (1978 - n + 1)}} + \cdots + \frac{1}{\sqrt{1978 \cdot 1}}\right) - 2 \cdot \frac{1978}{1979} > 0.$$

SOLUTION.

We can rewrite the inequality $\dfrac{a_1+a_2}{2} > \sqrt{a_1 \cdot a_2}$, where $a_1 \geq 0$, $a_2 \geq 0$, and $a_1 \neq a_2$ as equivalent inequality $\dfrac{1}{\sqrt{a_1 \cdot a_2}} > \dfrac{2}{a_1 + a_2}$, and use it 1978 times to each of the addends in the left-hand side of the given expression in parenthesis:

$$\frac{1}{\sqrt{1 \cdot 1978}} > \frac{2}{1+1978} = \frac{2}{1979},$$

$$\frac{1}{\sqrt{2 \cdot 1977}} > \frac{2}{2+1977} = \frac{2}{1979},$$

$$\cdots\cdots\cdots\cdots\cdots\cdots\cdots$$

$$\frac{1}{\sqrt{1978 \cdot 1}} > \frac{2}{1978+1} = \frac{2}{1979}.$$

Adding all these inequalities yields $\dfrac{1}{\sqrt{1 \cdot 1978}} + \dfrac{1}{\sqrt{2 \cdot 1977}} + \cdots + \dfrac{1}{\sqrt{1978 \cdot 1}} > 2 \cdot \dfrac{1978}{1979}.$

Therefore, this difference is indeed positive, i.e. $\left(\dfrac{1}{\sqrt{1 \cdot 1978}} + \dfrac{1}{\sqrt{2 \cdot 1977}} + \cdots + \dfrac{1}{\sqrt{1978 \cdot 1}} \right) - 2 \cdot \dfrac{1978}{1979} > 0$, which was required to be proved.

PROBLEM 9.5.

Which triangle out of all triangles with the given perimeter has the maximum area?

SOLUTION.

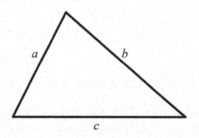

This problem concerns triangles with different sides' lengths and different angles' measures having one thing in common, they have the same perimeter. The question is which one of them has the greatest area. By definition, a perimeter of a triangle is the sum of its sides. Denoting the sides of an arbitrary triangle among our triangles a, b, and c, we have that its perimeter equals $P = a+b+c$. The formula relating the area of a triangle with its sides and its perimeter (more correct, with the semi-perimeter) is *Heron's formula* $S = \sqrt{p(p-a)(p-b)(p-c)}$, where p is the semi-perimeter of a triangle, that is, $p = \dfrac{P}{2} = \dfrac{a+b+c}{2}.$

Applying AM-GM inequality to three positive numbers $(p-a)$, $(p-b)$, and $(p-c)$, gives

$$\sqrt[3]{(p-a)(p-b)(p-c)} \leq \frac{(p-a)+(p-b)+(p-c)}{3}, \text{ or equivalently,}$$

$$(p-a)(p-b)(p-c) \leq \left(\frac{(p-a)+(p-b)+(p-c)}{3}\right)^3 = \left(\frac{p-a+p-b+p-c}{3}\right)^3 =$$

$$\left(\frac{3p-(a+b+c)}{3}\right)^3 = \left(\frac{3p-2p}{3}\right)^3 = \left(\frac{p}{3}\right)^3 = \frac{p^3}{27}.$$

Therefore,

$$S^2 = p(p-a)(p-b)(p-c) \leq p \cdot \frac{p^3}{27} = \frac{p^4}{27}, \text{ from which } S \leq \frac{p^2}{3\sqrt{3}} = \frac{p^2\sqrt{3}}{9} \text{ (*).}$$

So, by using AM-GM inequality, we established that the area of an arbitrary triangle does not exceed $\frac{p^2\sqrt{3}}{9}$, and it's not hard to notice that the maximum for S will be attained when triangle has equal sides, $a = b = c$. Indeed, for an equilateral triangle, its semi-perimeter equals $p = \frac{3}{2}a$. Hence, the area can be calculated as

$$S = \sqrt{p(p-a)(p-b)(p-c)} = \sqrt{\frac{3}{2}a \cdot \left(\frac{3}{2}a - a\right) \cdot \left(\frac{3}{2}a - a\right) \cdot \left(\frac{3}{2}a - a\right)} =$$

$$\sqrt{\frac{3}{2}a \cdot \frac{a}{2} \cdot \frac{a}{2} \cdot \frac{a}{2}} = \sqrt{\frac{3a^4}{16}} = \frac{a^2\sqrt{3}}{4}.$$

Expressing a in terms of p as $a = \frac{2}{3}p$ and substituting this value into the above equality,

yields $S = \frac{a^2\sqrt{3}}{4} = \frac{\left(\frac{2}{3}p\right)^2\sqrt{3}}{4} = \frac{p^2\sqrt{3}}{9}$. Comparing this result with (*), we see that this is the upper bound value we determined for the area of all triangles with the fixed given semi-perimeter p. It implies that the equilateral triangle will have the maximum area of all triangles with the same fixed perimeter.

Inequalities come in handy in solving many "maximum/minimum" calculation problems without referring to calculus means.

PROBLEM 9.6.

Given a cardboard sheet in a form of a square with the side a cut four congruent squares at its corners such that after bending them up an opened box of the greatest volume is made.

SOLUTION.

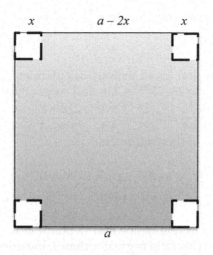

Denote the side of each of the four congruent squares by x. Then the volume of our open box in a form of a parallelepiped is $V = (a - 2x)^2 \cdot x$ (the product of the base area and the height). After multiplying both sides by 4, this can be rewritten as
$4V = (a - 2x) \cdot (a - 2x) \cdot 4x$. Now, we can apply AM-GM inequality to get that

$$4V = (a - 2x) \cdot (a - 2x) \cdot 4x \leq \left(\frac{a - 2x + a - 2x + 4x}{3} \right)^3 = \left(\frac{2a}{3} \right)^3.$$

Therefore, $V \leq \frac{1}{4} \cdot \left(\frac{2a}{3} \right)^3 = \frac{2a^3}{27}$. The equality is attained when $a - 2x = 4x$, i.e., when $x = \frac{a}{6}$.
Hence, the box of a maximum volume will be made if we cut four squares, each with the side of the length equal to $\frac{a}{6}$.

PROBLEM 9.7.

Find the minim value of the expression $(x + y)(x + z)$, knowing that $x > 0$, $y > 0$, $z > 0$, and $xyz(x + y + z) = 1$.

SOLUTION.

This problem has strikingly elegant and easy solution applying the AM-GM inequality.
Indeed, applying the AM-GM incquality, and the given condition that $xyz(x + y + z) = 1$, enables us to immediately observe that

$$(x + y)(x + z) = x^2 + zx + xy + zy = x(x + z + y) + zy \geq 2\sqrt{(x + z + y) \cdot xzy} = 2$$

which implies that the minimum value of our expression is 2.

In problems concerning comparison of two numbers finding the difference or ratio of them allows to get an easy result. If the difference is positive or ratio is greater than 1, then the first number is larger than the second one.

PROBLEM 9.8.

Decide which of these two numbers is larger: $\dfrac{23^{1981}+1}{23^{1982}+1}$ or $\dfrac{23^{1982}+1}{23^{1983}+1}$?

SOLUTION.

To keep our solution simple and avoid tedious calculations, let's introduce an auxiliary variable $n = 23^{1981}$. Then $23^{1982} = 23 \cdot 23^{1981} = 23n$ and respectively, $23^{1983} = 23^2 \cdot 23^{1981} = 23^2 n$.

It follows that $\dfrac{23^{1981}+1}{23^{1982}+1} = \dfrac{n+1}{23n+1}$ and $\dfrac{23^{1982}+1}{23^{1983}+1} = \dfrac{23n+1}{23^2 n+1}$.

Now, we will find the ratio of these numbers:

$$\frac{n+1}{23n+1} \div \frac{23n+1}{23^2 n+1} = \frac{n+1}{23n+1} \cdot \frac{23^2 n+1}{23n+1} = \frac{(23n)^2 + n(23^2 +1)+1}{(23n)^2 + 2 \cdot 23n+1} > 1.$$

Indeed, comparing addends in nominator and denominator of the last fraction, clearly, $n(23^2 +1) > 2 \cdot 23n$. Therefore this ratio is greater than 1, meaning that the first of the given numbers is larger than the second number, i.e., $\dfrac{23^{1981}+1}{23^{1982}+1} > \dfrac{23^{1982}+1}{23^{1983}+1}$.

PROBLEM 9.9.

Prove that for any positive numbers a and b, $\sqrt{\dfrac{a^2}{b}} + \sqrt{\dfrac{b^2}{a}} \geq \sqrt{a} + \sqrt{b}$.

PROOF.

Consider the difference $\sqrt{\dfrac{a^2}{b}} + \sqrt{\dfrac{b^2}{a}} - \left(\sqrt{a} + \sqrt{b}\right)$ and compare it to 0. If it is greater than or equal to 0, then our goal will be achieved, and the inequality is valid.

$$\sqrt{\frac{a^2}{b}} + \sqrt{\frac{b^2}{a}} - \sqrt{a} - \sqrt{b} = \frac{a}{\sqrt{b}} + \frac{b}{\sqrt{a}} - \sqrt{a} - \sqrt{b} = \frac{a\sqrt{a} + b\sqrt{b} - a\sqrt{b} - b\sqrt{a}}{\sqrt{ab}} =$$

$$\frac{a\left(\sqrt{a} - \sqrt{b}\right) - b\left(\sqrt{a} - \sqrt{b}\right)}{\sqrt{ab}} = \frac{(a-b)\left(\sqrt{a} - \sqrt{b}\right)}{\sqrt{ab}} =$$

$$\frac{\left(\sqrt{a} - \sqrt{b}\right)\left(\sqrt{a} + \sqrt{b}\right)\left(\sqrt{a} - \sqrt{b}\right)}{\sqrt{ab}} = \frac{\left(\sqrt{a} - \sqrt{b}\right)^2 \left(\sqrt{a} + \sqrt{b}\right)}{\sqrt{ab}} \geq 0.$$

The last fraction is greater or equal than 0 because $\left(\sqrt{a} - \sqrt{b}\right)^2 \geq 0$ as the square of a number, $\left(\sqrt{a} + \sqrt{b}\right) > 0$ as the sum of two positive numbers (by definition, square root of a number is a nonnegative number, and $\sqrt{ab} > 0$ as the square root of a positive number). This implies that the first number is larger than the second number and indeed, $\sqrt{\dfrac{a^2}{b}} + \sqrt{\dfrac{b^2}{a}} \geq \sqrt{a} + \sqrt{b}$,

as it was required to be proved.

Working with trigonometric inequalities, or comparisons of trigonometric expressions, we broadly utilize various trigonometric identities and properties of trigonometric functions.

PROBLEM 9.10.

Prove that $\sin^6 x + \cos^6 x \ge \dfrac{1}{4}$.

SOLUTION.

First, using the sum of cubes formula, $a^3 + b^3 = (a+b)(a^2 + b^2 - ab)$, and well-known trigonometric identities $\sin^2 x + \cos^2 x = 1$ and $\sin 2x = 2\sin x \cdot \cos x$, we modify the left-hand side as following:

$$\sin^6 x + \cos^6 x = \left(\sin^2 x\right)^3 + \left(\cos^2 x\right)^3 = \left(\sin^2 x + \cos^2 x\right)\left(\sin^4 x + \cos^4 x - \sin^2 x \cdot \cos^2 x\right) =$$

$$1 \cdot \left(\sin^4 x + \cos^4 x + 2\sin^2 x \cdot \cos^2 x - 3\sin^2 x \cdot \cos^2 x\right) = \left(\sin^2 x + \cos^2 x\right)^2 - 3(\sin x \cdot \cos x)^2 =$$

$$1 - 3 \cdot \frac{1}{4}\sin^2 2x.$$

Now, we will take the difference between the obtained result and $\dfrac{1}{4}$:

$$1 - 3 \cdot \frac{1}{4}\sin^2 2x - \frac{1}{4} = \frac{3}{4} - \frac{3}{4}\sin^2 2x = \frac{3}{4}\left(1 - \sin^2 2x\right) = \frac{3}{4}\cos^2 2x \ge 0.$$

The modified difference is greater than or equal 0, $\dfrac{3}{4}\cos^2 2x \ge 0$, because $\cos^2 2x \ge 0$, as the square of a number. It implies that indeed, $\sin^6 x + \cos^6 x \ge \dfrac{1}{4}$.

PROBLEM 9.11.

    a. Is it possible that $\sin x + \cos x = \sqrt[3]{3}$?

    b. Find angles x such that for $\tan x > 0$ and $\cot x > 0$, $\tan x + \cot x > 1.9$.

SOLUTIONS.

    a. Let's square both sides and investigate if this equality is possible,

$$(\sin x + \cos x)^2 = \left(\sqrt[3]{3}\right)^2.$$

It follows that, $\sin^2 x + \cos^2 x + 2\sin x \cdot \cos x = \sqrt[3]{9}$,

$$1 + \sin 2x = \sqrt[3]{9},$$

$\sin 2x = \sqrt[3]{9} - 1 > \sqrt[3]{8} - 1 = 2 - 1 = 1$. We arrive at conclusion that $\sin 2x > 1$, which is impossible for any real x because the range of the function sine is all real numbers not exceeding 1 in absolute value. Therefore, the equality in question is impossible.

b. $\tan x + \cot x = \tan x + \dfrac{1}{\tan x} \geq 2$ (because for any $a > 0,\ a + \dfrac{1}{a} \geq 2$). Hence

$\tan x + \cot x > 1.9$ for any real x, such that $\tan x > 0$, $\cot x > 0$.

PROBLEM 9.12.

Prove that $\cos \dfrac{x - y}{2} \geq \sqrt{\sin x \cdot \sin y}$.

SOLUTION.

First, notice that using the formula for cosine of the difference of two angles, we have

$$\cos \frac{x-y}{2} = \cos\left(\frac{x}{2} - \frac{y}{2}\right) = \cos\frac{x}{2}\cdot\cos\frac{y}{2} + \sin\frac{x}{2}\cdot\sin\frac{y}{2}.$$

Applying AM-GM inequality and the identity for the sine of a double-angle gives

$$\cos\frac{x-y}{2} = \cos\frac{x}{2}\cdot\cos\frac{y}{2} + \sin\frac{x}{2}\cdot\sin\frac{y}{2} \geq 2\sqrt{\cos\frac{x}{2}\cdot\cos\frac{y}{2}\cdot\sin\frac{x}{2}\cdot\sin\frac{y}{2}} =$$

$$\sqrt{\left(2\cos\frac{x}{2}\cdot\sin\frac{x}{2}\right)\cdot\left(2\cos\frac{y}{2}\cdot\sin\frac{y}{2}\right)} = \sqrt{\sin x \cdot \sin y},$$

as it was required to be proved.

As we've seen in previous chapters, trigonometry is often used as a supplemental tool for simplifying problems solutions. This proves true working with inequalities as well.

PROBLEM 9.13.

Given four positive numbers a, b, c, and d, prove that

$$\sqrt{ab} + \sqrt{cd} \leq \sqrt{(a+d)(b+c)}.$$

SOLUTION.

By definition, square root of a number is a positive number; so, we can divide both sides of inequality by $\sqrt{(a+d)(b+c)}$ without changing its sign to get $\dfrac{\sqrt{ab} + \sqrt{cd}}{\sqrt{(a+d)(b+c)}} \leq 1$, or

equivalently, $\sqrt{\dfrac{a}{a+d}\cdot\dfrac{b}{b+c}} + \sqrt{\dfrac{c}{b+c}\cdot\dfrac{d}{a+d}} \leq 1.$

For positive numbers $a, b, c,$ and d, clearly, each of the numbers $\dfrac{a}{a+d}, \dfrac{b}{b+c}, \dfrac{c}{b+c},$

and $\dfrac{d}{a+d}$ is less than 1, so there should exist angles β and γ $(0 < \beta, \gamma < \dfrac{\pi}{2})$ such that

$\dfrac{a}{a+d} = \sin^2\beta$ and $\dfrac{b}{b+c} = \sin^2\gamma$. Moreover, based on the trigonometric Pythagorean

Identity $\sin^2 x + \cos^2 x = 1$, since $\dfrac{a}{a+d} + \dfrac{d}{a+d} = \dfrac{a+d}{a+d} = 1,$ then $\dfrac{d}{a+d} = \cos^2\beta,$ and since

$\dfrac{b}{b+c} + \dfrac{c}{b+c} = \dfrac{b+c}{b+c} = 1,$ then $\dfrac{c}{b+c} = \cos^2\gamma.$

Now our assertion can easily be proved. It evolves from the formula for the cosine of the difference of two angles. Notice that using our substitutions we can rewrite the original inequality as

$$\sin\beta \cdot \sin\gamma + \cos\gamma \cdot \cos\beta \le 1.$$

Recalling the trigonometric identity for the cosine of the difference of two angles, we see that the left side of the last inequality is $\cos(\beta-\gamma)$. So, instead of proving the original inequality, we have to prove now that $\cos(\beta-\gamma) \le 1$. But this is an obvious fact. The range of the function cosine is all real numbers not exceeding 1 in absolute value. Therefore, $\cos(\beta-\gamma) \le 1$ for any angles β and γ $(0 < \beta, \gamma < \frac{\pi}{2})$. This implies that the original inequality is valid as well, and our proof is completed.

In proving many inequalities containing the second degree polynomials, properties of a quadratic function prove very useful and efficient.

A quadratic function in the variable x is defined by the formula $f(x) = ax^2 + bx + c$, where a, b, and c are real numbers $(a \ne 0)$. Recall that depending on a and the discriminant $D = b^2 - 4ac$, $ax^2 + bx + c > 0$ when $a > 0$, $D < 0$ and $ax^2 + bx + c < 0$ when $a < 0$, $D < 0$. The graph of the quadratic function, called parabola, is located either above or below X-axis. Graphically this can be depicted as the following:

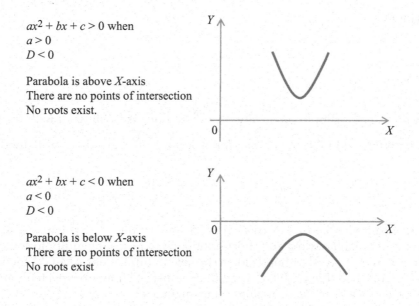

$ax^2 + bx + c > 0$ when
$a > 0$
$D < 0$

Parabola is above X-axis
There are no points of intersection
No roots exist.

$ax^2 + bx + c < 0$ when
$a < 0$
$D < 0$

Parabola is below X-axis
There are no points of intersection
No roots exist

PROBLEM 9.14.

Prove that $a^2 + b^2 + c^2 \ge ab + bc + ac$ for any real numbers a, b, and c.

SOLUTION.

Let's transfer all the terms on the left-hand side and rewrite this inequality as quadratic in a:

$$a^2 - a(b+c) + b^2 + c^2 - bc \ge 0. \ (*)$$

Finding the discriminant we get

$$D = (b+c)^2 - 4\left(b^2 + c^2 - bc\right) = b^2 + 2bc + c^2 - 4b^2 - 4c^2 + 4bc = -3b^2 - 3c^2 + 6bc =$$
$$- 3\left(b^2 + c^2 - 2bc\right) = -3(b-c)^2.$$

So, $D = -3(b-c)^2$ and this is clearly a non-positive number, $-3(b-c)^2 \leq 0$ (as the product of a negative number by a square of a number, which is always non-negative). The first coefficient by a^2 is 1, a positive number. Thus, the parabola will be located above X-axis, which means that indeed, our quadratic trinomial in a attains only positive values and it becomes 0, when $b = c$. Therefore, we conclude that $a^2 - a(b+c) + b^2 + c^2 - bc \geq 0$, or converting back to the original inequality, $a^2 + b^2 + c^2 \geq ab + bc + ac$, as it was required to be proved. Making one step further, let's investigate what happened when $b = c$. Substituting $b = c$ into (*) yields $a^2 - 2ab + b^2 + b^2 - b^2 \geq 0$, which is simplified to $(a-b)^2 \geq 0$. The equality to 0 is attained when $a = b$. So, we can now conclude that $a^2 + b^2 + c^2 \geq ab + bc + ac$ for any real numbers and the equality will be attained only when $a = b = c$.

Mathematical competitions often include problems in which one must prove an inequality in several variables. Various artificial tricks including even calculus applications are often useful in solving many Olympiad type challenges. Here are a few examples.

PROBLEM 9.15.

Prove that $x^3 + 8y^3 \geq 6xy^2$ for $x > 0$, $y > 0$.

<u>SOLUTION.</u>

Method 1.

Our inequality is equivalent to inequality $\dfrac{x^3 + 8y^3}{3} \geq 2xy^2$. We will work with the left-hand side and applying AM-GM Inequality for three positive numbers will modify it as following

$$\frac{x^3 + 8y^3}{3} = \frac{x^3 + 4y^3 + 4y^3}{3} \geq \sqrt[3]{x^3 \cdot 4y^3 \cdot 4y^3} = 2xy^2\sqrt[3]{2} > 2xy^2.$$

After multiplying both sides by 3, we obtain the desired result, $x^3 + 8y^3 \geq 6xy^2$. Even though this looks like a straightforward and short solution, it requires ingenuity to come up with the idea how to modify the left-hand side to be able to implement AM-GM Inequality. In our second approach to this problem, we will demonstrate how unexpectedly useful calculus techniques may be in tackling difficult inequalities.

Method 2.

Since it is given that $x > 0$, we can divide both sides of inequality by x^3 without changing its sign and get equivalent inequality $1 + 8\left(\dfrac{y}{x}\right)^3 \geq 6\left(\dfrac{y}{x}\right)^2$. Denoting $u = \dfrac{y}{x}$, and slightly modifying our inequality, we now have to prove that $8u^3 - 6u^2 + 1 \geq 0$ for $u > 0$ (clearly, since $x > 0$ and $y > 0$, then $u = \dfrac{y}{x} > 0$).

Let's consider the function $f(u) = 8u^3 - 6u^2 + 1$ and find its derivative

$$f'(u) = 24u^2 - 12u.$$

$f'(u) > 0$ when $24u^2 - 12u > 0$, or equivalently, when $u(2u-1) > 0$.

Solving last inequality, we see that our continuous function is increasing on the intervals $]-\infty, 0[\cup \left]\frac{1}{2}, +\infty\right[$ and it is decreasing on the interval $\left]0, \frac{1}{2}\right[$ ($f'(u) < 0$ on this interval).

Hence, function $f(u) = 8u^3 - 6u^2 + 1$ has its minimum at point $u = \frac{1}{2}$. It implies, that for any

$u > 0$, $f(u) > f\left(\frac{1}{2}\right) = 8 \cdot \frac{1}{8} - 6 \cdot \frac{1}{4} + 1 = \frac{1}{2}$. Therefore, we see that $8u^3 - 6u^2 + 1 > \frac{1}{2} > 0$. Turning

back to the original inequality, this means that $\dfrac{x^3 + 8y^3}{3} \geq 2xy^2$, which concludes our proof.

PROBLEM 9.16.
Compare e^π and π^e.

SOLUTION.
Let's first do some preliminary work and notice that instead of comparing the given two numbers, after taking natural logarithm of each number, we may compare the numbers $(\pi \ln e)$ and $(e \ln \pi)$. Moreover, by dividing each number by the positive number $(\ln e \cdot \ln \pi)$, we can compare the numbers $\dfrac{\pi}{\ln \pi}$ and $\dfrac{e}{\ln e}$. Analyzing which one of these two numbers is greater, we will arrive at comparison of the original numbers.

Drawing an analogy with the technique applied in Method 2 in the previous problem, we will consider the function $f(x) = \dfrac{x}{\ln x}$, $x > 0$, $x \neq 1$.

Let's find the derivative of this function and examine its behavior. We will apply the quotient rule for the derivative of a fraction, $\left(\dfrac{u}{v}\right)' = \dfrac{u' \cdot v - v' \cdot u}{v^2}$.

So, $f'(x) = \dfrac{1 \cdot \ln x - x \cdot \dfrac{1}{x}}{\ln^2 x} = \dfrac{\ln x - 1}{\ln^2 x}$. Now, we will solve inequality $f'(x) > 0$, i.e., $\dfrac{\ln x - 1}{\ln^2 x} > 0$,

and determine the intervals where our function increases. Since $\ln^2 x > 0$ as the square of a number, our inequality reduces to $\ln x - 1 > 0$. The solutions of this inequality are all numbers from the interval $]e, +\infty[$. In other words, our $f'(x) > 0$ for all numbers such that $x > e$. So, function $f(x)$ is continuous and increasing for $x > e$, which implies that since $\pi > e$ then $f(\pi) = \dfrac{\pi}{\ln \pi} > f(e) = \dfrac{e}{\ln e}$. Respectively, going in a reverse order from the last inequality to the original one, we finally conclude that

$$e^\pi > \pi^e.$$

One of the classic inequalities that is very useful in many difficult problems is *Cauchy-Bunyakovsky inequality* (after French mathematician Augustin-Louis Cauchy (1789–1857) and the Ukrainian mathematician Viktor Bunyakovsky (1804–1889)). It states that for any real numbers the following inequality holds true:

$$\left(a_1 b_1 + a_2 b_2 + \cdots + a_n b_n\right)^2 \le \left(a_1^2 + a_2^2 + \cdots + a_n^2\right)\left(b_1^2 + b_2^2 + \cdots + b_n^2\right).$$

The equality holds only when numbers a_k and b_k are the multiple of each other.

There are several proofs of this inequality. We will demonstrate an elegant non-conventional proof applying vector properties techniques.

PROOF.

Let's consider two vectors defined by their coordinates in n-dimensional space $\vec{a} = (a_1, a_2, \ldots, a_n)$ and $\vec{b} = (b_1, b_2, \ldots, b_n)$. The dot product of these two vectors through their coordinates is determined as

$$\vec{a} \cdot \vec{b} = a_1 b_1 + a_2 b_2 + \cdots + a_n b_n. \tag{9.1}$$

On the other hand, we know that the same dot product can be calculated as the product of the length of $\vec{a}$ by the length of $\vec{b}$ and by the cosine of the angle between them:

$$\vec{a} \cdot \vec{b} = \|\vec{a}\| \cdot \|\vec{b}\| \cdot \cos \gamma. \tag{9.2}$$

Comparing (9.1) and (9.2) and recalling that for any angle γ, $|\cos \gamma| \le 1$ we get that

$$a_1 b_1 + a_2 b_2 + \cdots + a_n b_n = \|\vec{a}\| \cdot \|\vec{b}\| \cdot \cos \gamma =$$

$$\sqrt{a_1^2 + a_2^2 + \cdots + a_n^2} \cdot \sqrt{b_1^2 + b_2^2 + \cdots + b_n^2} \cdot \cos \gamma \le \sqrt{a_1^2 + a_2^2 + \cdots + a_n^2} \cdot \sqrt{b_1^2 + b_2^2 + \cdots + b_n^2}.$$

Squaring both sides of the last inequality, we will obtain the sought-after result

$$\left(a_1 b_1 + a_2 b_2 + \cdots + a_n b_n\right)^2 \le \left(a_1^2 + a_2^2 + \cdots + a_n^2\right)\left(b_1^2 + b_2^2 + \cdots + b_n^2\right).$$

The equality will be achieved when $\dfrac{a_1}{b_1} = \dfrac{a_2}{b_2} = \cdots = \dfrac{a_n}{b_n}$, and we hope that readers can easily justify this assertion independently.

PROBLEM 9.17.

What is the maximum of the function $f(x) = a \sin x + b \cos x$, where $a > 0, b > 0$, and $0 < x < \dfrac{\pi}{2}$?

SOLUTION.

Applying Cauchy-Bunyakovsky inequality and recalling trigonometric Pythagorean Identity that $\sin^2 x + \cos^2 x = 1$ gives

$$f(x) = a \sin x + b \cos x \le \sqrt{a^2 + b^2} \cdot \sqrt{\sin^2 x + \cos^2 x} = \sqrt{a^2 + b^2}.$$

The maximum that is equal to $\sqrt{a^2 + b^2}$ will be attained when $\dfrac{a}{b} = \dfrac{\sin x}{\cos x}$, i.e., when $\tan x = \dfrac{a}{b}$.

PROBLEM 9.18.

Given that $a + b + c = 1$, prove that $a^2 + b^2 + c^2 \geq \dfrac{1}{3}$.

<u>SOLUTION.</u>

Applying Cauchy-Bunyakovsky inequality gives

$$1 = a \cdot 1 + b \cdot 1 + c \cdot 1 \leq \sqrt{a^2 + b^2 + c^2} \cdot \sqrt{1^2 + 1^2 + 1^2} = \sqrt{3} \cdot \sqrt{a^2 + b^2 + c^2}.$$

Therefore, indeed, after dividing both sides by $\sqrt{3}$ and squaring, we obtain the desired result, $a^2 + b^2 + c^2 \geq \dfrac{1}{3}$. It is interesting to note that in case when $a = b = c = \dfrac{1}{3}$ then $a + b + c = 1$, and we obtain the exact equality $a^2 + b^2 + c^2 = \dfrac{1}{3}$.

Inequalities are broadly utilized in solving non-standard equations and systems of equations, as it is demonstrated in the following Olympiad type problems 9.19 and 9.20.

PROBLEM 9.19.

Prove that for any natural n ($n \in N$, $n \neq 0$) there exists the unique set of numbers $x_1, x_2, \ldots, x_n$ such that these numbers satisfy the equation

$$(1 - x_1)^2 + (x_1 - x_2)^2 + \cdots + (x_{n-1} - x_n)^2 + x_n^2 = \frac{1}{n+1}.$$

<u>SOLUTION.</u>

Amazingly, this difficult Olympiad type problem has an elegant and vivid solution which evolves from one of the classic inequalities considered in Chapter 4, namely, the inequality relating the arithmetic mean and quadratic mean,

$$\frac{a_1 + a_2 + \cdots + a_n}{n} \leq \sqrt{\frac{a_1^2 + a_2^2 + \cdots + a_n^2}{n}}.$$

After squaring both sides this inequality can be rewritten as

$$\frac{a_1^2 + a_2^2 + \cdots + a_n^2}{n} \geq \left(\frac{a_1 + a_2 + \cdots + a_n}{n} \right)^2$$

or equivalently, after multiplying by n,

$$a_1^2 + a_2^2 + \cdots + a_n^2 \geq \frac{(a_1 + a_2 + \cdots + a_n)^2}{n}.$$

Setting $a_1 = 1 - x_1$, $a_2 = x_1 - x_2$, ..., $a_{n-1} = x_{n-1} - x_n$, $a_n = x_n$, we get that

$$(1 - x_1)^2 + (x_1 - x_2)^2 + \cdots + (x_{n-1} - x_n)^2 + x_n^2 \geq \frac{1}{n+1} \left((1 - x_1) + (x_1 - x_2) + \cdots + (x_{n-1} - x_n) + x_n \right)^2 =$$

$$\frac{1}{n+1} (1 - x_1 + x_1 - x_2 + \cdots + x_{n-1} - x_n + x_n)^2 = \frac{1}{n+1}.$$

The equality will be achieved when $1 - x_1 = x_1 - x_2 = \cdots = x_{n-1} - x_n = x_n = \dfrac{1}{n+1}$.

So,

$$x_1 = 1 - \frac{1}{n+1},$$

$$x_2 = x_1 - \frac{1}{n+1} = 1 - \frac{1}{n+1} - \frac{1}{n+1} = 1 - \frac{2}{n+1},$$

$$x_3 = x_2 - \frac{1}{n+1} = 1 - \frac{2}{n+1} - \frac{1}{n+1} = 1 - \frac{3}{n+1},$$

....

$$x_{n-1} = 1 - \frac{n-1}{n+1},$$

i.e., the equality is achieved when $x_i = 1 - \dfrac{i}{n+1}$ for ($i = 1, 2, \ldots, n-1$). We indeed obtained the unique set of numbers $x_1, x_2, \ldots, x_n$ expressed in terms of $n \in N$ (for any natural n) as described above, such that they satisfy our equation. Our proof is complete.

PROBLEM 9.20.

Find positive solutions of the system of equations

$$\begin{cases} x_1 + x_2 + x_3 + \cdots + x_n = 3, \\ \dfrac{1}{x_1} + \dfrac{1}{x_2} + \dfrac{1}{x_3} + \cdots + \dfrac{1}{x_n} = 3. \end{cases}$$

SOLUTION.

Assume we managed to find the set of numbers $x_1, x_2, x_3, \ldots, x_n$ satisfying the given system of the equations. Let's denote $A = \sqrt[n]{x_1 \cdot x_2 \cdot \ldots \cdot x_n}$.

Applying the AM – GM inequality to numbers $x_1, x_2, x_3, \ldots, x_n$ gives

$$x_1 + x_2 + x_3 + \cdots + x_n \geq n \cdot \sqrt[n]{x_1 \cdot x_2 \cdot \ldots \cdot x_n} = n \cdot A.$$

Since it is given in the first equation of the system that $x_1 + x_2 + x_3 + \cdots + x_n = 3$, we obtain that $n \cdot A \leq 3$, from which

$$A \leq \frac{3}{n}. \tag{9.3}$$

Applying now the AM – GM inequality to numbers $\dfrac{1}{x_1}, \dfrac{1}{x_2}, \ldots, \dfrac{1}{x_n}$, we get

$$\frac{1}{x_1} + \frac{1}{x_2} + \frac{1}{x_3} + \cdots + \frac{1}{x_n} \geq n \cdot \sqrt[n]{\frac{1}{x_1} \cdot \frac{1}{x_2} \cdot \ldots \cdot \frac{1}{x_n}} = \frac{n}{\sqrt[n]{x_1 \cdot x_2 \cdot \ldots \cdot x_n}} = \frac{n}{A}.$$

It is given in the second equation of the system that $\dfrac{1}{x_1} + \dfrac{1}{x_2} + \dfrac{1}{x_3} + \cdots + \dfrac{1}{x_n} = 3$. Therefore, $\dfrac{n}{A} \le 3$, from which

$$A \ge \frac{n}{3}. \tag{9.4}$$

Comparing (9.3) and (9.4) yields $\dfrac{n}{3} \le A \le \dfrac{3}{n}$. Multiplying both sides by a natural number n, $n > 0$, we obtain that $\dfrac{n^2}{3} \le A \cdot n \le 3$, which implies the inequality $n^2 \le 9$. Obviously, for natural n to satisfy the last inequality, it has to be either 1, 2, or 3 $(n \ne 0)$.

For $n = 1$ the system has no solutions ($x_1 = 3$ and $\dfrac{1}{x_1} = 3$, which is impossible).

If $n = 2$, we get the system of equations

$$\begin{cases} x_1 + x_2 = 3, \\ \dfrac{1}{x_1} + \dfrac{1}{x_2} = 3. \end{cases}$$

Simplifying the second equation and substituting the value of $x_1 + x_2 = 3$ gives $\dfrac{x_2 + x_1}{x_1 \cdot x_2} = 3$ or equivalently, $x_1 \cdot x_2 = 1$. Hence, our system converts to

$$\begin{cases} x_1 + x_2 = 3, \\ x_1 \cdot x_2 = 1. \end{cases}$$

By Viète's theorem, x_1 and x_2 have to be the roots of the quadratic equation

$$y^2 - 3y + 1 = 0.$$
$$D = 9 - 4 = 5.$$
$$y = \frac{3 \pm \sqrt{5}}{2}.$$

Both roots are positive numbers. Therefore, there are two pairs of the positive solutions of the original system for $n = 2$, $\left(\dfrac{3+\sqrt{5}}{2}, \dfrac{3-\sqrt{5}}{2} \right)$ and $\left(\dfrac{3-\sqrt{5}}{2}, \dfrac{3+\sqrt{5}}{2} \right)$.

Finally, if $n = 3$, we have

$$\begin{cases} x_1 + x_2 + x_3 = 3, \\ \dfrac{1}{x_1} + \dfrac{1}{x_2} + \dfrac{1}{x_3} = 3. \end{cases}$$

Inequality $\dfrac{n^2}{3} \le A \cdot n \le 3$ can be rewritten as $3 \le 3A \le 3$ or equivalently, $1 \le A \le 1$. Therefore, for $n = 3$, we have $1 \le \sqrt[3]{x_1 \cdot x_2 \cdot x_3} \le 1$. This implies that $\sqrt[3]{x_1 \cdot x_2 \cdot x_3} = 1$. The last equation has the only positive solutions $x_1 = x_2 = x_3 = 1$ satisfying the given system of the equations.

Answer: when $n = 1$ or $n \geq 4$, there are no positive solutions;

- when $n = 2$, there are two solutions $\left(\dfrac{3+\sqrt{5}}{2}, \dfrac{3-\sqrt{5}}{2} \right)$ and $\left(\dfrac{3-\sqrt{5}}{2}, \dfrac{3+\sqrt{5}}{2} \right)$;

- when $n = 3$, there is one solution $(1, 1, 1)$.

In the remainder of this chapter we will focus on geometrical inequalities. This topic is perhaps one of the most genuine forms of problem solving requiring deep knowledge and understanding of links between algebra, geometry, and trigonometry. Many problems involving geometrical inequalities present great examples of charm and beauty of close relationships in various math disciplines.

PROBLEM 9.21.

Prove that product of any two sides of a triangle is not less than product of its perimeter by radius of its incircle.

SOLUTION.

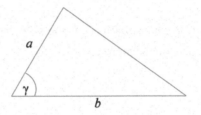

To get to the desired result it suffices to recall two formulas for calculating the area of a triangle, one expressing the area in terms of two sides and sine of an angle between them, $S_\Delta = \dfrac{1}{2} ab \cdot \sin \gamma$, and the second formula expressing the same area in terms of its perimeter and radius of its incircle, $S_\Delta = \dfrac{1}{2} Pr$ (we've referred to this formula several times in previous chapters). Comparing these formulas, we see that $ab \cdot \sin \gamma = Pr$. But since $|\sin \gamma| \leq 1$ for any real values of γ, than clearly, $ab \geq Pr$, and we are done!

We can make one interesting observation from this problem. If the angle between sides a and b is not a right angle, then $ab > Pr$, and the following inequality holds for any triangle with sides a, b, and c:

$$ab + bc + ac > 3Pr.$$

PROBLEM 9.22.

Given the lengths of the sides of a convex quadrilateral a, b, c, and d (going in a clockwise direction), prove that the area of the quadrilateral is not greater than

$$\frac{1}{4}(a+b)(c+d).$$

SOLUTION.

First, let's simplify the given expression $\frac{1}{4}(a+b)(c+d)$ and see if we can derive some helpful ideas from its different presentation:

$$\frac{1}{4}(a+b)(c+d) = \frac{1}{4}(ad+bc) + \frac{1}{4}(ac+bd).$$

By drawing the diagonal BD we cut our quadrilateral into two triangles DAB and BCD.

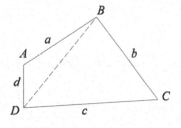

FIGURE 9.1

Recalling the formula for calculating the area of a triangle as half the product of its sides by sine of the angle between them and the fact that $|\sin\alpha| \leq 1$ for any angle α, we have

$$S_{DAB} = \frac{1}{2}ad \cdot \sin A \leq \frac{1}{2}ad,$$

$$S_{BCD} = \frac{1}{2}bc \cdot \sin C \leq \frac{1}{2}bc.$$

Adding the areas of triangles DAB and BCD we get the area of $ABCD$. So, we see that

$$S_{ABCD} = S_{DAB} + S_{BCD} \leq \frac{1}{2}ad + \frac{1}{2}bc = \frac{1}{2}(ad+bc). \tag{9.5}$$

Next, we will cut again our quadrilateral across its diagonal BD, but now we will flip the sides BC and CD. We will end up getting the quadrilateral congruent to the original one (see Figure 9.2).

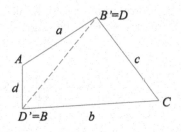

FIGURE 9.2

Now, we will draw the diagonal AC (see Figure 9.3) and in the same fashion as we did above, calculate the area of our quadrilateral as the sum of the areas of the other pair of triangles, $AB'C$ and $CD'A$, and make the similar conclusion regarding its upper bound:

$$S_{ABCD} = S_{AB'C} + S_{CD'A} \leq \frac{1}{2}ac + \frac{1}{2}bd = \frac{1}{2}(ac+bd). \tag{9.6}$$

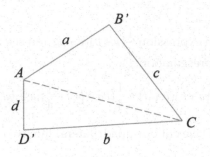

FIGURE 9.3

Dividing both sides of (9.5) and (9.6) by 2 and adding them gives the sought-after result

$$S_{ABCD} \le \frac{1}{4}(ad + bc) + \frac{1}{4}(ac + bd) = \frac{1}{4}(a + b)(c + d).$$

It is noteworthy that applying the above results and using the fact that for positive real numbers x and y it follows that $\frac{x}{y} + \frac{y}{x} \ge 2$ (or equivalently, $xy \le \frac{x^2 + y^2}{2}$) we easily can get another geometric inequality relating the area of a convex quadrilateral with the sum of the squares of its sides:

$$S_{ABCD} \le \frac{a^2 + b^2 + c^2 + d^2}{4}.$$

This inequality immediately follows from (9.6) observing that $ac \le \frac{a^2 + c^2}{2}$ and $bd \le \frac{b^2 + d^2}{2}$ and adding these two inequalities.

PROBLEM 9.23.

This problem was offered by V. Senderov in magazine Квант (in Russian) #1, 2000 – M1692.
    Given that a, b, and c are the numbers expressing lengths of the sides of a triangle, prove that

$$\frac{a^2 + 2bc}{b^2 + c^2} + \frac{b^2 + 2ac}{a^2 + c^2} + \frac{c^2 + 2ab}{a^2 + b^2} > 3.$$

SOLUTION.

The hint to easy and nice solution is hidden in the given conditions about a, b, and c that they represent the lengths of the sides of a triangle. Recalling the Triangle inequality theorem we know that $a + c > b$, or equivalently, $a > b - c$. Squaring both sides of the last inequality gives $a^2 > (b - c)^2$, which can be rewritten as $a^2 + 2bc > b^2 + c^2$. Clearly, $b^2 + c^2 > 0$ as the sum of two positive numbers, therefore, we can divide by this number both sides of the last inequality preserving its sign: $\dfrac{a^2 + 2bc}{b^2 + c^2} > 1$. In the same way we obtain that $\dfrac{b^2 + 2ac}{a^2 + c^2} > 1$

and $\dfrac{c^2 + 2ab}{a^2 + b^2} > 1$. Adding these three inequalities gives the desired result

$$\frac{a^2 + 2bc}{b^2 + c^2} + \frac{b^2 + 2ac}{a^2 + c^2} + \frac{c^2 + 2ab}{a^2 + b^2} > 3.$$

PROBLEM 9.24.

Prove that for any triangle with sides a and b and the angle bisector l_c of the angle between these sides the following inequality holds:

$$l_c < \sqrt{ab}.$$

PROOF.

There are many alternative proofs of this inequality. We will demonstrate a pure geometrical solution. We will also apply earlier proved trigonometrical inequality from Problem 9.12.

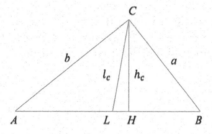

In triangle ABC we draw the angle bisector of angle ACB, $CL = l_c$ and the altitude CH to the side AB, $CH = h_c$. We have $BC = a$ and $AC = b$. For convenience, let's also denote $\angle CAB = \alpha$ and $\angle CBA = \beta$.

Our goal will be to examine right triangle CHL and get several useful relationships between h_c and l_c allowing us to tackle the requested inequality.

First, let's express angle LCH in terms of α and β.

Observe, that in triangle ABC, $\angle ACB = 180° - (\alpha + \beta)$, and since CL divides this angle in half,

$$\angle ACL = \frac{1}{2}\left(180° - (\alpha + \beta)\right) = 90° - \frac{\alpha + \beta}{2}. \tag{9.7}$$

Also, in a right triangle AHC,

$$\angle ACH = 90° - \alpha. \tag{9.8}$$

Now the stage is set for expressing angle LCH as $\angle LCH = \angle ACH - \angle ACL$.

Substituting (9.7) and (9.8) into the above equality gives $\angle LCH = 90° - \alpha - \left(90° - \dfrac{\alpha + \beta}{2}\right) = \dfrac{\alpha + \beta}{2} - \alpha = \dfrac{\beta - \alpha}{2}$ (we assume that $\beta > \alpha$, as it is depicted in our diagram).

Next, we turn to a right triangle CHL ($\angle LHC = 90°$), and find that $l_c = \dfrac{h_c}{\cos\left(\dfrac{\beta - \alpha}{2}\right)}$.

Recall now that by the extended law of sines applied to triangle ABC, $\dfrac{a}{\sin\alpha} = \dfrac{b}{\sin\beta} = 2R$

(where R is the radius of the circumscribed circle), from which $a = 2R \cdot \sin\alpha$ and $b = 2R \cdot \sin\beta$.

Considering a right triangle BHC and substituting $BC = a$ from the above expression we get $h_c = CH = BC \cdot \sin\beta = a \cdot \sin\beta = 2R \cdot \sin\alpha \cdot \sin\beta$.

Finally, applying trigonometric inequality $\cos\dfrac{x-y}{2} \geq \sqrt{\sin x \cdot \sin y}$ that we proved in Problem 9.12, we have

$$l_c = \frac{h_c}{\cos\left(\dfrac{\beta - \alpha}{2}\right)} < \frac{h_c}{\sqrt{\sin\alpha \cdot \sin\beta}} = \frac{2R \cdot \sin\alpha \cdot \sin\beta}{\sqrt{\sin\alpha \cdot \sin\beta}} = 2R\sqrt{\sin\alpha \cdot \sin\beta} =$$

$$\sqrt{(2R \cdot \sin\alpha) \cdot (2R \cdot \sin\beta)} = \sqrt{ab}.$$

We get that $l_c < \sqrt{ab}$, which is the result we need, and our proof is completed.

It merits to point out that while applying the result obtained in Problem 9.12 for evaluating l_c in our final step, we used a "<" sign not a "≤" sign. Referring back to AM-GM inequality that we applied in Problem 9.12, we know that the equality is attained only

when $\cos\dfrac{x}{2} \cdot \cos\dfrac{y}{2} = \sin\dfrac{x}{2} \cdot \sin\dfrac{y}{2}$, which can be written as $\dfrac{\cos\dfrac{x}{2}}{\sin\dfrac{x}{2}} = \dfrac{\sin\dfrac{y}{2}}{\cos\dfrac{y}{2}}$, or equivalently,

$\cot\dfrac{x}{2} = \tan\dfrac{y}{2}$. Since $\cot\dfrac{x}{2} = \tan\left(90° - \dfrac{x}{2}\right)$, the last equality is possible only when $90° - \dfrac{x}{2} = \dfrac{y}{2}$.

This implies that $x + y = 180°$, which is impossible for two angles in a triangle. This brief analysis justifies *"less than"* instead of *"less than or equal"* sign.

One final remark – the readers who familiar with the property of an angle bisector that $l_c^2 = ab - AL \cdot BL$ (this property was mentioned at the conclusion of Chapter 5) can save the effort and get strikingly simple proof of our inequality:

$$l_c^2 = ab - AL \cdot BL < ab.$$

Simple as that! In many instances knowledge of formulas relating various elements in a triangle proves invaluable in a problem-solving process.

PROBLEM 9.25.

Prove that for any right triangle with legs a and b and hypotenuse c,

$$a + b \leq c\sqrt{2}.$$

In this classic final problem of the chapter we will demonstrate three different solutions, the first – purely geometrical, the second – applying algebraic techniques, and the third one – applying trigonometric identities.

PROOF 1.

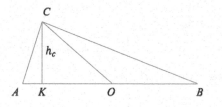

Consider $\triangle ABC$ ($\angle C = 90°$) in which $BC = a$, $AC = b$, and $AB = c$.

We drop the altitude CK on the hypotenuse AB, and denote it $CK = h_c$. Next, we express the area of triangle ABC in two different ways, as half the product of its legs, and as half the product of the altitude by the hypotenuse:

$$S_{ABC} = \frac{1}{2}ab \text{ and } S_{ABC} = \frac{1}{2}ch_c.$$

Therefore,

$$ab = ch_c \tag{9.9}$$

Also, note that by the Pythagorean theorem,

$$a^2 + b^2 = c^2 \tag{9.10}$$

It's easy to show that in any right triangle, $h_c \leq \frac{c}{2}$. Indeed, denoting O the midpoint of AB, we can consider another right triangle CKO ($\angle K = 90°$) in which obviously, leg CK is not longer than the hypotenuse CO. Noticing that $CO = AO = OB = \frac{c}{2}$ (O is the center of the circumcircle of triangle ABC), we get

$$h_c \leq \frac{c}{2}. \tag{9.11}$$

Finally, let's do a few simple algebraic manipulations expanding $(a+b)^2$ and substituting values from (9.9) and (9.10), and using the inequality (9.11):

$$(a+b)^2 = a^2 + b^2 + 2ab = c^2 + 2ch_c \leq c^2 + 2c \cdot \frac{c}{2} = c^2 + c^2 = 2c^2.$$

Since a, b, and c are positive numbers, from the last expression it follows that $a + b \leq c\sqrt{2}$, and we are done.

PROOF 2.

Let's use the obvious algebraic identity $(a+b)^2 + (a-b)^2 = 2(a^2 + b^2)$.

Since $a^2 + b^2 = c^2$ and $(a-b)^2 \geq 0$ (a square of a number is always non-negative), we have $(a+b)^2 \leq 2(a^2 + b^2) = 2c^2$. The last inequality implies that for positive a, b, and c, $a + b \leq c\sqrt{2}$, which is the desired result.

PROOF 3.

Expressing the legs in the right triangle ABC through the hypotenuse and trigonometric functions, we get $a = c \cdot \sin A$ and $b = c \cdot \cos A$.

Then, recalling that $\sin 45° = \cos 45° = \dfrac{1}{\sqrt{2}}$ (we will multiply and divide by this number) and $\sin \alpha \cdot \cos \beta + \sin \beta \cdot \cos \alpha = \sin(\alpha + \beta)$, we have

$$a + b = c \cdot \sin A + c \cdot \cos A = c\sqrt{2}\left(\frac{1}{\sqrt{2}}\sin A + \frac{1}{\sqrt{2}}\cos A\right) =$$

$$c\sqrt{2}\left(\cos 45° \cdot \sin A + \sin 45° \cdot \cos A\right) = c\sqrt{2} \cdot \sin(45° + A).$$

We know that $|\sin x| \leq 1$ for any real number x. Therefore, $|\sin(45° + A)| \leq 1$ and it follows that $a + b = c\sqrt{2} \cdot \sin(45° + A) \leq c\sqrt{2} \cdot 1 = c\sqrt{2}$, and our proof is completed.

There is a feature inherent in almost any beautiful mathematical result: many paths lead to it. As we just demonstrated, we can approach interesting geometrical inequalities from different sides, and all the paths enchant those who eager to take them. We encourage ambitious readers to further explore and investigate different solutions and compare them. Then you should be able to select the one you find appealing, instructive, and the most interesting for you.

10

Guess and Check Game

In a sense, mathematics has been most advanced by those who distinguished themselves by intuition rather than by rigorous proofs.

Felix Klein

When learning science, you usually try to explain what you've observed. One starts with an educated guess and, by testing it, one begins to establish a more elaborate framework, some theory or a model. What about mathematics?

There are many different ways to approach mathematical problems; among them there is a strategy that doesn't sound overly mathematical, solving problems using guess and check. But it can be very useful when done properly. While it is certainly true that mathematicians do not consider as a good idea to simply guess a random answer for a problem, there are instances when educated guesses are helpful and useful. A good estimate is not just a random guess. *Accurate* estimate might give one a real boost in finding a proper plan for solving a math problem. Besides, solving problems using guess and check is a process that requires logic and a clear understanding of the question. In many instances, it may help not just to make the solution manageable, but it may significantly simplify the solution process. Tackling difficult problems, one may be at sea where to start contemplating a plan for a solution. The educated guess (hypothesis) might become the first step in a solution process. Furthermore, even the incorrect guess or assumption may shed light on some important properties not seen before. Surely, for any successful guess, rigid proof is always required.

We invite the readers to solve the considered below problems by applying conventional techniques. You can compare then your solutions with the suggestions below and decide for yourself which strategy is more appealing in these cases.

PROBLEM 10.1.

The exact dates of birth and death of the prominent ancient Greek mathematician Diophantus are unknown. However, the legend states that the following epitaph is written on his grave:

> "Here lays Diophantus, the wonders behold.
> Through art algebraic, the stone tells how old:
> God gave him his boyhood one-sixth of his life,
> One twelfth more as youth while whiskers grew rife;
> And then yet one-seventh ere marriage begun;
> In five years there came a bouncing new son.
> Alas, the dear child of master and sage
> After attaining half the measure of his father's life
> chill fate took him. After consoling his fate by the
> science of numbers for four years, he ended his life."

How old was he when he died?

DOI: 10.1201/9781003359500-10

SOLUTION.

Algebraic method reduces this problem to solving an equation, which is the most natural and obvious way of solving this problem. However, there exists another strategy for attacking this problem. Since we are looking for an integer solution (our concern is about his age expressed obviously as some natural number), then it has to be a multiple of 6, 12, and 7. The least common multiple of 6, 12, and 7 is 84. Logically, it makes no sense to consider higher common multiples. Hence, 84 is the desired answer. How easy and efficient was this solution?!

The next two problems are taken from the great collection of Sam Loyd's problems.

PROBLEM 10.2.

"Together these two turkeys weigh twenty pounds," said the butcher. "The little fellow sells for 2 cents a pound more than the big bird".
   Mrs. Smith bought the little one for 82 cents and Mrs. Brown paid $2.96 for the big turkey. How much did each gobbler weigh?

SOLUTION.

Let's see if we can use a guess-and-check strategy to simplify the solution. The price paid for the big turkey is almost 4 times of what was paid for the little one, because $2.96 \div 0.82 \approx 3.61$. In many cases, there is a slight discount for similar purchases of bigger items compared to the price of smaller items. So, we may reasonably assume that the weight of the little turkey should be $\frac{1}{4}$ of the weight of the big one. Knowing that together they weigh 20 pounds, it is not hard to figure out their weights as 4 and 16 pounds respectively. Then, in our assumptions, the cost per pound of the little turkey is $82 \div 4 = 20.5$ cents, and the cost per pound of the big turkey is $296 \div 16 = 18.5$ cents. We see that the difference in cost per pound between them is $20.5 - 18.5 = 2$ cents. This implies that the conditions of the problem are satisfied and our guess was correct. The weight of the small turkey was 4 pounds and the weight of the big one was 16 pounds.

PROBLEM 10.3.

A lady gave the postage stamp clerk a one-dollar bill and said, "Give me some two-cent stamps, ten times as many one-cent stamps, and the balance in fives." How can the clerk fulfil this puzzling request?

SOLUTION.

Observe that the number of 2-cent stamps must be divisible by 5; otherwise, the total number of all stamps that she requested from clerk will not satisfy the conditions of the problem to be sold for exactly 1 dollar. She wanted to have one-cent stamps 10 times more than two-cent stamps. Hence, the number of one-cent stamps should be divisible by $5 \cdot 10 = 50$. Clearly, then there should be exactly 50 one-cent stamps for the total price of 50 cents. It implies that there will be 5 two-cent stamps for the total price of 10 cents and the price of the rest of the stamps, five-cent stamps, has to be $100 - 10 - 50 = 40$ cents. So, there should be $40 \div 5 = 8$ five-cent stamps.
   Let's now take another view at this problem.
   Applying algebraic approach, we may denote the requested number of two-cent stamps by x, and the number of five-cent stamps by y. The number of one-cent stamps is 10 times

the number of two-cent cents, so it is $10x$. It follows that the problem is reduced to solving the equation $2x + 10x + 5y = 100$. Now you can clearly see that since $10x$, $5y$, and 100, each is divisible by 5, then the first addend of $2x$ on the left-hand side of the equation has to be divisible by 5 as well. This enlightens our observation at the beginning of the solution (if for some readers that fact was not obvious). The last equation simplifies to $12x + 5y = 100$, and it is the so-called Diophantine equation with two variables that is solvable in integers. Clearly, the only possible natural solution for x is 5, implying that $y = \dfrac{(100 - 12 \cdot 5)}{5} = 8$.

PROBLEM 10.4.

A student purchased several similar notebooks and several same letter-size color folders. He paid $10.88 in total for the notebooks. How many notebooks did he buy if the price of each notebook exceeds more than by $1 the price of a color folder, and he purchased six more notebooks than folders?

SOLUTION.

The conditions of the problem imply that student purchased not less than 7 notebooks (it is known that he purchased 6 more notebooks than folders). Also, since the price of each notebook exceeds more than by $1 the price of a color folder, then each notebook has to cost more than $1. Since the total price paid for notebooks was $10.88, it is clear that he purchased not more than 10 notebooks. So, we reduced the possible number of purchased notebooks to 7, 8, 9, or 10. Out of these four numbers only 8 is a factor of 10.88 providing the cost of each notebook equal to $1.36. Therefore, we arrive at conclusion that he purchased 8 notebooks.

In tackling certain equations the most elegant, easy, and efficient way of solving is to guess the root and prove that there are no other solutions. We will start with problem B196 offered in *Quantum*, March/April 1997 issue.

PROBLEM 10.5.

Solve the equation:

$$10 - 9\left(9 - 8\left(8 - 7\left(7 - 6\left(6 - 5\left(5 - 4\left(4 - 3\left(3 - 2(2 - x)\right)\right)\right)\right)\right)\right)\right) = x.$$

SOLUTION.

You can expand out and eventually find that $x = 1$. But much easier is to guess that $x = 1$ (I would say, this is a natural guess) and check this solution. Since the equation is linear, it cannot have more than one solution, and we are done!

PROBLEM 10.6.

Solve the equation $(3x + 5)\sqrt{4x - 1} = 11\sqrt{7}$.

SOLUTION.

First, determining the domain of this equation, we see that $x > \dfrac{1}{4}$ since the expression under the square root cannot be negative and $x \neq \dfrac{1}{4}$ because otherwise left-hand side becomes equal to 0 and $0 \neq 11\sqrt{7}$. The way this equation is written, may lead to an assumption that the factors with radical signs on both sides should be equal. Then the solution of our equation is reduced to solving the following system of equations:

$$\begin{cases} 3x+5=11, \\ \sqrt{4x-1}=\sqrt{7}. \end{cases}$$

This system simplifies to

$$\begin{cases} 3x=6, \\ 4x=8. \end{cases}$$

Solving each equation, we obtain the common solution $x=2$. Let's prove now that the original equation has no other solutions.

Indeed, if $x>2$, then obviously, $(3x+5)\sqrt{4x-1}>(3\cdot 2+5)\cdot\sqrt{4\cdot 2-1}=11\sqrt{7}$.

If $\frac{1}{4}<x<2$, then $(3x+5)\sqrt{4x-1}<11\sqrt{7}$. So, in either case, we see that there are no other solutions. Therefore, 2 is the only root of our equation.

The conventional approach to solve the given equation is to get rid of irrationality on the left-hand side of the equation by squaring both sides, and then continue solving the cubic equation. It translates into pretty tedious and lengthy process comparing to the suggested above strategy utilizing an intelligent guess.

Answer: $x=2$.

PROBLEM 10.7.

Solve the equation $1+4^{2x}=5^{2x}$.

SOLUTION.

Dividing both sides by $5^{2x}\left(5^{2x}\neq 0\right)$ yields $\left(\frac{1}{5}\right)^{2x}+\left(\frac{4}{5}\right)^{2x}=1$ (*).

It's easy to see that $x=\frac{1}{2}$ satisfies the equation, $\left(\frac{1}{5}\right)^{1}+\left(\frac{4}{5}\right)^{1}=1$ – true statement. Relying on the properties of exponential function, let's prove that there are no other solutions.

Indeed, if $x>\frac{1}{2}$, then $\left(\frac{1}{5}\right)^{2x}+\left(\frac{4}{5}\right)^{2x}<\frac{1}{5}+\frac{4}{5}=1$, which contradicts to (*).

If $x<\frac{1}{2}$, then $\left(\frac{1}{5}\right)^{2x}+\left(\frac{4}{5}\right)^{2x}>\frac{1}{5}+\frac{4}{5}=1$, which contradicts to (*).

Therefore, $x=\frac{1}{2}$ is the only root of the original equation.

Answer: $x=\frac{1}{2}$.

PROBLEM 10.8.

Solve the equation $\sqrt{x-3} \cdot \sqrt{5-x} \cdot \sqrt{x-4} = -x^2 + 8x - 16$.

<u>SOLUTION</u>.

Let's first slightly modify the equation by completing a square on the right-hand side. Using the formula $a^2 - 2ab + b^2 = (a-b)^2$, we can rewrite the equation as

$$\sqrt{x-3} \cdot \sqrt{5-x} \cdot \sqrt{x-4} = -(x-4)^2.$$

Before going any further, we should find the permissible domain for x. It is determined from the solutions of the following inequalities:

$$\begin{cases} x-3 \geq 0, \\ 5-x \geq 0, \\ x-4 \geq 0. \end{cases}$$

It follows that

$$\begin{cases} x \geq 3, \\ x \leq 5, \\ x \geq 4. \end{cases}$$

All real values of x from the segment $[4, 5]$ satisfy our system of inequalities.

We can easily see that $x = 4$ is the solution of our equation. Indeed, $\underbrace{\sqrt{4-3} \cdot \sqrt{5-4} \cdot \sqrt{4-4}}_{0} = \underbrace{-(4-4)^2}_{0}$, and 4 belongs to the domain $[4, 5]$ of our equation. So, we guessed the root, and now our task is to investigate if there are any more solutions. Clearly, there are none, because the product of three non-negative numbers on the left-hand side (square root of a number n by definition is always a nonnegative number, $\sqrt{n} \geq 0$) is a non-negative number, whilst there is a non-positive number on the right-hand side of the equation $(-(x-4)^2 \leq 0$, since $(x-4)^2 \geq 0$ as the square of some number). It implies that the equality is attained only for such values of x when both sides equal 0. Therefore, $x = 4$ is the unique solution of the equation. This completes our solution.

Answer: $x = 4$.

PROBLEM 10.9.

Solve the equation $\sin^5 x + \cos^3 x = 1$.

SOLUTION.

Equation looks tough and most likely attempts to solve it applying conventional techniques are bound to fail. However, just by looking at this equation it is easy to guess its roots, values of x for which $\sin x = 1$ and $\cos x = 0$ or $\cos x = 1$ and $\sin x = 0$. The solutions are $x = \dfrac{\pi}{2} + 2\pi k$, $k \epsilon Z$ and $x = 2\pi n$, $n \epsilon Z$. Let's now prove that there are no other solutions.

Clearly, for any values of x other than mentioned above, $\sin x \neq 0$, $\sin x \neq 1$, $\cos x \neq 0$, $\cos x \neq 1$; hence, recalling the properties of exponential function a^x for $0 < a < 1$ (and we know that $|\sin x| < 1$) we get that $\sin^5 x < \sin^2 x$ and $\cos^3 x < \cos^2 x$.

Applying the trigonometric Pythagorean Identity, it follows that

$$\sin^5 x + \cos^3 x < \sin^2 x + \cos^2 x = 1.$$

The last observation proves that no other solutions exist.

Answer: $x = \dfrac{\pi}{2} + 2\pi k$, $k \epsilon Z$ and $x = 2\pi n$, $n \epsilon Z$.

It is impossible to specify a universal method for making good guesses.

However, since solving geometrical problems we usually heavily rely on diagrams depicting the data of the problem, an accurate drawing can help us make a good guess. In some instances, an educated guess might be enlightening helping to either find a direct solution or convert the original problem to a simpler and easily manageable problem.

We have to be extra cautious, though, every time making an assumption based on the drawing alone. On all such occasions, the rigid proof of the validity of our assumption is required.

PROBLEM 10.10.

There are given acute angled triangle ABC and a random straight line l passing through its centroid M. Knowing that the distances to l from vertices A, B, and C are x, y, and z respectively, prove that $x = y + z$.

SOLUTION.

It is given that M is the centroid of $\triangle ABC$, $AK \perp l$, $AK = x$, $CF \perp l$, $CF = y$, $BN \perp l$, $BN = z$. We need to prove that $x = y + z$.

Even after closely examining the diagram, it is not clear how we can relate the segments in question, AK, CF, and BN by their lengths. What we may notice, though, that by drawing the perpendiculars from B and C to l we formed the trapezoid $CFNB$. Bearing this thought in mind, let's do a few auxiliary constructions. In triangle ABC we draw the median AE to the side BC; it contains M, as it is given to us that M is the centroid of $\triangle ABC$, and next, we draw $ED \perp l$. Since E is the midpoint of CB, then by the *Thale's theorem* (it is also known as *Intercept theorem*), D is the midpoint of FN. Thus, DE is the midline in trapezoid $CFNB$, and it can be expressed in terms of its bases as $DE = \dfrac{y+z}{2}$.

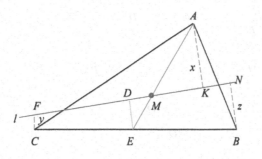

Our goal is to prove that $x = AK = y + z$. Should we assume that $DE = \frac{1}{2} AK$? By the way, just looking at our figure, we may get the same feeling as well. If this assumption is correct and we will be able to prove it, the problem will be solved. Let's see how valid our guess is. Being a perpendicular to the same straight line l, $DE \parallel AK$. Checking angles, we conclude that triangles EDM and AKM are similar ($\angle EDM = \angle AKM = 90°$ and $\angle EMD = \angle AMK$, as vertical). It follows that $\dfrac{DE}{AK} = \dfrac{EM}{AM}$. Recalling now that M is the centroid, so it divides AE in the ratio 2: 1, we get that $\dfrac{DE}{AK} = \dfrac{EM}{AM} = \dfrac{1}{2}$. So, indeed, $DE = \frac{1}{2} AK$. Substituting the values in terms of x, y, and z, we arrive at the conclusion that $x = y + z$. Our assumption turned out to be correct. It allowed us to shift the focus from the sum of two segments in question, CF and BN, to an auxiliary segment DE and relate its length to AK. This significantly simplified our path to the desired result, which, by the way, is an amazing aspect of the centroid's property:

for any straight line l passing through the triangle's centroid and separating vertices B and C from A, the sum of perpendiculars dropped from B and C to l equals to the length of the perpendicular from A to l.

The proved relationship is valid for any triangle because we did not consider a specific line at M; line l was any randomly drawn line through the centroid of the triangle.

Problem 10.11 presents a simplified version of the problem examined in the final chapter of my book "The Equations World", Dover Publications, 2019. It was solved by applying the Cartesian coordinate system technique. The goal there was to demonstrate the benefits of algebraic technique in setting the system of equations while tackling geometrical problems. Here we will exhibit a pure geometrical solution. Good reasonable guess is instrumental in transforming our original problem to a different problem which happens to be much easier to solve. In fact, the educated guess in this example plays a huge role in making the solution not just manageable but elegant as well.

PROBLEM 10.11.

There is given the right angle and a segment with end points lying on the sides of the angle. The segment moves in such a way that its length remains fixed and its end points slide on the sides of the given angle. With every move a square is built on the segment as on its side inside of the given angle. Find the locus of the centers of the squares.

SOLUTION.

We are given the right angle EKT and segment AB ($A \in KE$, $B \in KT$). The length of AB stays fixed while its end points A and B slide on KE and KT. As AB moves, various squares are built on AB inside the angle EKT. The question is what geometrical figure is formed by the centers of all such squares.

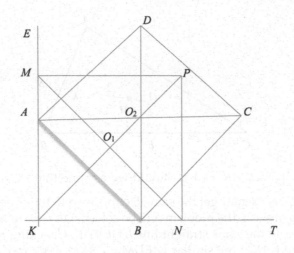

FIGURE 10.1

Let's first examine the two cases when AB is fully located on KE (B coincides with K and A with M, such that $AB = MK$) and AB is fully located on KT (B coincides with N and A with K, such that $AB = KN$). These are the two "extreme" positions of AB in its sliding on the sides of the given angle.

The "extreme" squares built on AB, i.e. on KM and KN, will coincide and will be presented by the square $KMPN$ (see Figure 10.1). Looking for the center of this square, let's draw its diagonals MN and KP, and denote O_1 the point of their intersection.

Second, let's now build a square on AB as on its side when AB takes the position such that $AB \parallel MN$, the square $ABCD$. Denote O_2 the point of intersection of its diagonals AC and BD. By our constructions, O_1 and O_2 lie on the bisector of the given angle EKT (justify this!). Should we assume that for any other location of AB the center of a square built on AB as on its side will lie on the angle bisector of EKT as well? This looks like a reasonable hypothesis. Let's verify if it is indeed an accurate assertion.

We now draw an arbitrary segment AB (its length stays the same) with A on KE and B on KT, different from the three already considered configurations. To make the diagram easily readable, we will not show the full square built on AB but show in Figure 10.2 only the point of intersection of its diagonals O_3 (this is the point of interest).

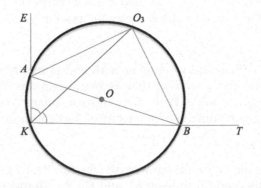

FIGURE 10.2

This point is such that $AO_3 = BO_3$ and $\angle AO_3B = 90°$ (the diagonals of a square bisect each other and are perpendicular). Notice that in the newly obtained quadrilateral KAO_3B we have a pair of right opposite angles, $\angle AO_3B$ and $\angle AKB$. Therefore, KAO_3B is a cyclic quadrilateral (the opposite angles of a cyclic quadrilateral are supplementary), that is, there exists a circumcircle of KAO_3B with the center at the midpoint of AB, point O, and radius $r = \frac{1}{2}AB$.

Remember, our assumption was that O_3 has to lie on the angle bisector of EKT. So, our goal now is to prove (or disprove; the negative result is also a useful result!) that KO_3 is the angle bisector of AKB (this is the same angle as EKT).

Now, our auxiliary circle enlightens the whole picture. We see that $\angle AKO_3 = \angle BKO_3$ as inscribed angles subtended by the equal chords AO_3 and BO_3 ($AO_3 = BO_3$ as halves of the diagonals of a square). We, therefore, arrive at the conclusion that indeed KO_3 is the angle bisector of AKB, which means that O_3 lies on the angle bisector of angle EKT. Because this time we considered an arbitrary location of AB inside the given angle, our statement is valid for center of any square built on AB inside of angle EKT, and we can conclude that all such centers lie on the angle bisector of EKT. Analyzing our diagram in Figure 10.1, we should make the second reasonable assumption that the centers of all such squares will form the segment between O_1 and O_2. In other words, the locus of all such centers in question, per this assumption, should be the segment O_1O_2.

Let's denote a the fixed length of AB, $AB = a$. It follows that $KO_1 = \frac{1}{2}KP = \frac{a\sqrt{2}}{2}$ (half of the diagonal of the square $KMPN$) and $KO_2 = AB = a$ (as the diagonals of the square KAO_2B). Our goal will be to prove that for arbitrary location of AB, other than parallel to MN and coinciding with MK, the center of a square built on AB, point O_3, is such that it lies between O_1 and O_2, that is, $KO_1 < KO_3 < KO_2$, i.e. $\frac{a\sqrt{2}}{2} < KO_3 < a$.

Clearly, being a chord in the circle with the diameter $AB = a$, $KO_3 < a$.

To prove the second part of our assertion that $KO_3 > \frac{a\sqrt{2}}{2}$, we will refer to Figure 10.2, where we considered the case of an arbitrary location of AB inside of our angle EKT.

Consider acute-angled $\triangle KO_3B$. We proved that $\angle O_3KB = \frac{1}{2} \cdot 90° = 45°$. On the other hand, $\angle O_3BK > 45°$, because $\angle O_3BK = \angle O_3BA + \angle KBA = 45° + \angle KBA$ ($\angle O_3BA = 45°$ as the angle by the base of the right isosceles $\triangle AO_3B$). As we know, a greater angle of a triangle is opposite a greater side; hence, $KO_3 > O_3B$. Now, we merely have to observe that in right isosceles $\triangle AO_3B$, $O_3B = O_3A = \frac{a\sqrt{2}}{2}$. Therefore, indeed,

$$KO_3 > \frac{a\sqrt{2}}{2},$$

and we see that our second assumption happens to be correct as well. This completes our proof that the locus of all the centers of the squares built on the moving segment AB inside EKT is the segment connecting O_1 and O_2 contained on the angle bisector of EKT.

In our solution above, intelligent hypothesis helped to transform the original problem into an equivalent but much simpler problem. The introduction of an auxiliary circle illuminated all our next steps in the solution process. Moreover, it allowed us to get an easy proof of our second assumption regarding the location of centers lying on the angle bisector of EKT between O_1 and O_2. These two assumptions were instrumental in getting relatively short and nice solution.

PROBLEM 10.12.

There is given a triangle ABC with sides $AB = 4$, $AC = \sqrt{17}$, and $BC = 5$. Point D lies on AB such that $AD = 1$. Find the distance between the centers of the circumcircles of the triangles BDC and ADC.

SOLUTION.

This problem looks tough. The accurate diagram here should provide some hints to contemplate the plan for the solution. By examining the diagram, it looks like under the given conditions, CD is an altitude dropped to AB. Is this a valid guess?

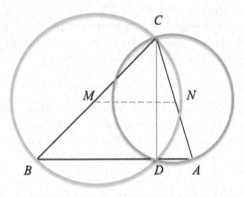

How this assumption can be verified? Well, we can use the given lengths of the sides to observe that $BC^2 - BD^2 = AC^2 - AD^2$.

Indeed, $5^2 - (4-1)^2 = 16$ and $\left(\sqrt{17}\right)^2 - 1^2 = 16$.

Applying the theorem converse to the Pythagorean Theorem, we see that CD is the common leg of two right triangles BDC and ADC. Surprisingly, the final outcome almost immediately evolves from the perpendicularity of CD and AB. We know that the center of the circumcircle of a right triangle is the midpoint of the hypotenuse. Hence, M, the midpoint of BC, is the center of the circumscribed circle of $\triangle BDC$, and N, the midpoint of AC, is the center of the circumscribed circle of $\triangle ADC$. But this means that the length of MN is the sought-after distance in question. It remains only to observe that MN is the midline in triangle ABC, and thus, $MN = \dfrac{1}{2} AB = 2$, the distance we seek. Our assumption about CD being perpendicular to AB happened to be correct and it indeed significantly simplified our solution.

An elegant trick that can sometimes be used to solve geometric problems is to "substitute" certain elements under consideration into other elements (angles, sides, or even entire figures) that are in some way more convenient. The impact of an intelligent guess in those instances should not be underestimated.

PROBLEM 10.13.

Let C be the point chosen outside of an angle AOB, $\angle AOB = 60°$. Three perpendiculars CD, CM, and CN are dropped at C to OA, OB, and to the angle bisector ON of angle AOB respectively. Find ON knowing that $CM = a$ and $CD = b$.

<u>**SOLUTION.**</u>

Let's denote the points of intersection of CN (and its extension) with OA and OB by P and E respectively. Analyzing the diagram, the natural desire is to assume that the newly formed triangle OPE is equilateral. If the latter is true, we can draw $PK \perp CM$ and $PF \perp OE$, and instead of finding the length of ON, we can find the length of PF as the congruent angle bisector and altitude (at the same time!) in the equilateral triangle OPE.

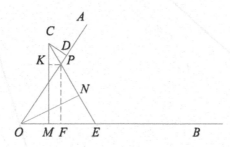

This can be easily done because PF should be equal to the difference of CM and CK (if we manage to prove that OPE is an equilateral triangle, then it will not be hard to also prove that $CK = CD = b$) and we will get that $PF = KM = CM - CK = a - b$.

Our analysis gives us the first glimmer of an idea and the hope of success. So, let's see if we can justify our assumption regarding $\triangle OPE$ to be an equilateral triangle.

We know that $CN \perp ON$ and that $\angle PON = \angle EON = 60° : 2 = 30°$ (because it is given that ON is the angle bisector of $\angle AOB = \angle POE$). Therefore, in right triangle ONP, $\angle OPN = 90° - 30° = 60°$. It implies that there are two $60°$ angles ($\angle POE$ and $\angle OPE$) in isosceles triangle OPE which indeed makes it equilateral. Now, we observe that $\angle OPE = \angle CPD = 60°$ as vertical angles. Also, $\angle CPK = \angle CEO = 60°$ as corresponding angles by parallel lines OE and KP (they are parallel because by construction, each is perpendicular to CM) and transversal CE. We obtained that in two right triangles CDP and CKP all the angles are congruent. Noting that they have the common hypotenuse CP, we conclude that these triangles are congruent by Angle-Side-Angle property. It follows that the corresponding legs will be congruent as well, which justifies our second assumption that $CK = CD = b$.

In equilateral triangle OPE all its altitudes, medians, and angle bisectors dropped to each of three sides are congruent. Recall now that we drew $PF \perp OE$. It implies that being the altitude in equilateral triangle OPE, $PE = ON$. Finally, at this point we can appreciate our original assumptions and observe that instead of determining the length of ON we can substitute it for the congruent segment PF and find its length. By construction, $KPFM$ is a rectangle, and therefore, $PF = KM$. Hence, $ON = PF = KM = CM - CK = a - b$, and we are done.

In our next cute geometrical puzzle, intelligent guessing "game" is essential in making solution vivid and easy to follow.

PROBLEM 10.14.

Three squares are built on the sides of a right triangle ACB ($\angle C = 90°$). Connect the vertices of these squares to form three triangles MAD, FBE, and NCK. Prove that these three triangles have equal areas.

SOLUTION.

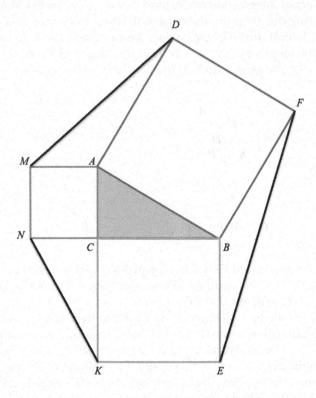

As we see in a diagram above, three squares $ACNM$, $ADFB$, and $CBEK$ are constructed on the sides of the right triangle ACB. Our goal is to prove that the areas of the formed triangles MAD, FBE, and NCK are equal.

Since we are working with the squares, clearly, two pairs of sides in each of the considered triangles are congruent to two respective sides in our right triangle:

In $\triangle NCK$ and $\triangle ACB$, $NC = AC$ and $CK = CB$.

In $\triangle MAD$ and $\triangle ACB$, $MA = AC$ and $AD = AB$.

In $\triangle FBE$ and $\triangle ACB$, $FB = AB$ and $BE = CB$.

First, we immediately observe that right triangles $\triangle NCK$ and $\triangle ACB$ are congruent (they have equal legs), so they have equal areas. We know that the area of a triangle can be calculated as half the product of the base by the altitude dropped to that base. Should we assume that the altitudes dropped to the respective bases in each of the remaining two triangles will be of the same length?

By doing the additional constructions of the altitudes in our triangles we clearly see that our assumption is correct (see figure below). Amazingly, we now can easily arrive at conclusion that the base of each triangle $\triangle MAD$ and $\triangle FBE$ and the altitude dropped to that base are respectively congruent to the legs of our "basic" right triangle $\triangle ACB$. It implies that the area of each of these triangles is equal to the area of our right triangle! And consequently, three triangles MAD, NCK, and FBE have equal areas. We hope that it should be a good exercise for the readers to complete the rigid proof of our assertions.

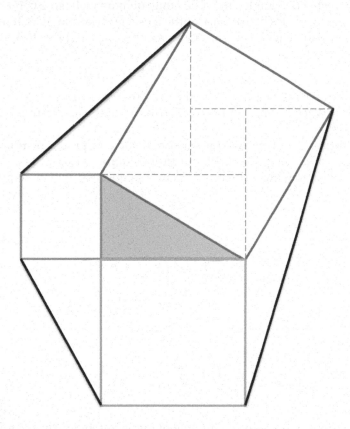

In conclusion of our geometrical explorations with guessing game when we heavily relied on diagrams involved, we will offer an entertaining geometrical sophism. It demonstrates how careful one has to be in his/her assumptions based on a diagram alone, and how critical it is to provide a rigid proof of one's hypothesis.

PROBLEM 10.15.

Try to find an error in the following justification of the fact that a right angle can be equal to an obtuse angle.

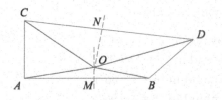

We draw a segment AB and $AC \perp AB$. Next, we draw $BD = AC$ such that $\angle ABD > 90°$.

Clearly, AB and CD are not parallel, therefore, the perpendicular bisectors to each segment must intersect at some point O, $MO \perp AB$ (M is the midpoint of AB), $NO \perp CD$ (N is the midpoint of CD), and $MO \cap NO = O$. It follows that $AO = BO$ and $CO = DO$ because O is equidistant respectively from A and B, and from C and D. Observing that by construction, $BD = AC$, we obtain that triangles AOC and BOD are congruent by three sides. Then their respective angles must be congruent as well, so $\angle CAO = \angle DBO$. Also, $\angle OAB = \angle OBA$, as the angles by the base of the isosceles triangle AOB. It implies that the sums of equal angles are equal as well, $\angle CAO + \angle OAB = \angle DBO + \angle OBA$, i.e.

$$\angle CAB = \angle DBA$$

But the last equality means that the right angle $\angle CAB$ equals to the obtuse angle $\angle DBA$. How come is this possible?! (The solution can be found in the appendix).

In solving certain types of problems, the most realistic and practical strategy is to guess the correct outcome and then prove that your proposition is valid.

Let's consider the following arithmetic "gem", solution of which requires logic and intuition.

PROBLEM 10.16.

Given that

$$2 + 3 = 10$$
$$7 + 2 = 63$$
$$6 + 5 = 66$$
$$8 + 4 = 96,$$

Determine what is

$$9 + 7 = ?$$

<u>SOLUTION</u>.

This problem is entirely about making intelligent guesses! Obviously, the given conditions have nothing to do with straight forward addition in decimal system we use for counting. It is some type of an arithmetic cryptograph with a secret message which we need to decipher. If we will be able to find the common pattern in all four equalities, the same pattern can be applied to the sum in question, "9 + 7", to figure out what it has to represent.

Working out various scenarios (we hope the readers will try the problem, before looking our explanations), we note that if we divide the result by the sum in the first equality, it gives the first addend, $10 : (2 + 3) = 2$. Is our guess correct and we can apply this pattern to solve the problem? Let's verify our assumption for the other equalities.

Indeed, $63 : (7 + 2) = 7; 66 : (6 + 5) = 6; 96 : (8 + 4) = 8$.

So, we obtained the following results:

$10 : 5 = 2$ (2 is the first addend in "$2 + 3 = 10$")

$63 : 9 = 7$ (7 is the first addend in "$7 + 2 = 63$")

$66 : 11 = 6$ (6 is the first addend in "$6 + 5 = 66$")

$96 : 12 = 8$ (8 is the first addend in "$8 + 4 = 96$").

Therefore, $9 + 7 = 144$ because $144 : (9 + 7) = 144 : 16 = 9$. The problem is solved.

Don't we have a very similar approach facing problems related to infinite sequences that require determining an explicit rule (it gives its n-th term as a function of the term's position number n in the sequence) given its recurrence rule (when one or more initial terms are given and each further term of a sequence is defined as a function of the preceding terms)? One could guess (of course, it is an *intelligent guess*!) the explicit rule by considering few first terms in a sequence and then checking the formula. The next step has to be validating one's assumption and finding the rigid proof for all the terms by applying, for instance, mathematical induction.

PROBLEM 10.17.

Find the explicit rule expressing the n-th term of the sequence of natural numbers such that $x_n = 5x_{n-1} - 6x_{n-2}$, where $x_1 = 1$, $x_2 = 2$.

SOLUTION.

Analyzing the first five terms after x_2, we see that

$$x_3 = 5 \cdot 2 - 6 \cdot 1 = 4,$$

$$x_4 = 5 \cdot 4 - 6 \cdot 2 = 8,$$

$$x_5 = 5 \cdot 8 - 6 \cdot 4 = 16,$$

$$x_6 = 5 \cdot 16 - 6 \cdot 8 = 32,$$

$$x_7 = 5 \cdot 32 - 6 \cdot 16 = 64.$$

You may have noticed that all our calculations resulted in powers of 2.

Thus, one can reasonably guess that the formula in question for the n-th term is $x_n = 2^{n-1}$, when $n \geq 3$.

The technique of mathematical induction naturally springs to mind.

First, we verify our proposition for $n = 3$.

If $n = 3$, then $x_3 = 4 = 2^{3-1}$, and we see that our assumption holds.

Assume now, that for $n = k - 1$, $x_{k-1} = 2^{k-2}$ and for $n = k$, $x_k = 2^{k-1}$. We need to prove that in these assumptions, $x_{k+1} = 2^k$.

Consider

$$x_{k+1} = 5x_k - 6x_{k-1} = 5 \cdot 2^{k-1} - 6 \cdot 2^{k-2} = 5 \cdot 2^{k-1} - \left(5 \cdot 2^{k-2} + 1 \cdot 2^{k-2}\right) = 5 \cdot 2^{k-1} - 5 \cdot 2^{k-2} - 2^{k-2} =$$
$$5 \cdot 2^{k-2}(2-1) - 2^{k-2} = 5 \cdot 2^{k-2} - 2^{k-2} = 4 \cdot 2^{k-2} = 2^2 \cdot 2^{k-2} = 2^{2+k-2} = 2^k,$$

which is what we wanted to prove. So, we see that our guess was correct and the explicit formula for the nth term in our sequence is $x_n = 2^{n-1}$ for $n \geq 3$.

As we concentrated in this chapter on practical applications of educated "guess-and-check" strategy for tackling math problems and emphasized its relevance in honing one's mathematical skills, we provided our readers with another handy tool to add in their math toolbox. In my mind, one of the most important things a person can get studying mathematics (for anyone, not just for future professional mathematicians) is the attainment of a higher intellectual level. In solving problems we not only learn how to prove what's true, we also learn on many occasions how to guess at the truth. And the ability to guess is a part of effective thinking.

In conclusion, one more remark - it is important to stress out that this method is closely related to exploring analogies in mathematical disciplines. Analogies play an important role in mathematics. Through analogy at times, we can come to notable conclusions without tedious calculations. The intuition we develop in solving math problems applying "guess-and-check" strategy can help us use the powerful tool of analogy properly. It is worth mentioning here so-called "Fermi problems" (after prominent Italian physicist Enrico Fermi (1901–1954)), the estimation technique to make good approximate calculations with little or no actual data. Ambitious readers can fill in the interest in this subject doing further research independently. I can refer to one very interesting article concerning this topic, "Educated Guesses" by John A. Adam, published in *Quantum*, September/October 1995 issue.

This is the last chapter in the book. And to make our final thoughts fun, in conclusion we offer one intriguing logical hypothesis leading to a very interesting idea that has nothing to do with mathematics. Obviously, the suggested technique considering and verifying an intelligent proposition often is applicable not just in mathematics, but in real life situations as well.

In many languages word "night" consists of letter n and a word either directly or very closely resembling a digit 8:

In English: *night = n + eight (8)*

In French: *nuit = n + huit (8)*

In German: *nacht = n + acht (8)*

In Italian: *notte = n + oito (8)*

In Portuguese: *noite = n + oito (8)*

In Spanish: *noche = n + ocho (8)*

Should these examples lead us to assume that this is a hint about how many hours a normal night should last? In other words, should the healthiest sleep last 8 hours? Amazingly, we know from medical studies that most healthy adults need between 7 to 8 hours of sleep per night to function at their best! What do you think?

Do you have any similar real-life observations from your experience?

> *Mathematics has a threefold purpose. It must provide an instrument for the study of nature. But this is not all: it has a philosophical purpose, and, I daresay, an aesthetic purpose.*

> **Henri Poincare**

Appendix

For Chapter 1, Problem 1.12

To prove that if the inscribed angle in a circle is subtended by a diameter, then it is a right angle.

This assertion is a corollary from so-called *The inscribed angle theorem* (also known as *Star Trek Lemma*) which relates the measure of an inscribed angle to that of the central angle subtending the same arc:

An angle inscribed in a circle is half of the central angle that subtends the same arc on the circle.

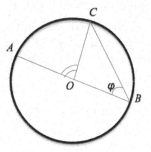

Let O be the center of a circle. Pick two random points B and C on the circle, and connect three of them to form an isosceles ΔBOC ($BO = CO$ as radii). Extend BO till its intersection with circle at A. We need to prove that the inscribed angle ABC is half of the central angle AOC.

$\angle OCB = \angle OBC = \varphi$ as the angles by the base of the isosceles triangle BOC. Since angle AOC is exterior angle of the triangle BOC, $\angle AOC = \angle OCB + \angle OBC = 2\varphi$, which proves the validity of our assertion, an angle inscribed in a circle is half of the central angle that subtends the same arc on the circle.

For any inscribed angle subtended by the diameter, its value is $180° \cdot \dfrac{1}{2} = 90°$. Hence, we see that any such angle is a right angle.

For Chapter 2, Page 23

Geometric mean theorem or the Right triangle altitude theorem states that the geometric mean of the two segments that the altitude in a right triangle on the hypotenuse creates on the hypotenuse equals the altitude.

Direct Theorem:
Given: ΔACB, where $\angle ACB = 90°$, $CH \perp AB$.
To prove: $HC = \sqrt{AH \cdot BH}$.

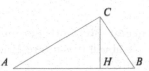

Proof.
Checking angles, we see that all three triangles $\triangle ACB$, $\triangle AHC$, and $\triangle CHB$ are similar, $\triangle ACB \sim \triangle AHC \sim \triangle CHB$. From the ratios of the respective sides, we easily get the desired result:

$$\frac{AH}{HC} = \frac{HC}{HB}, \text{ from which } HC = \sqrt{AH \cdot BH}.$$

Converse Theorem (use the same diagram as above):
Given: $\triangle ACB$, where, $CH \perp AB$ and $HC = \sqrt{AH \cdot BH}$.
To prove: $\angle ACB = 90°$.

Proof.
First, we rearrange the given equality $HC = \sqrt{AH \cdot BH}$ as $\dfrac{AH}{HC} = \dfrac{HC}{HB}$. Next, observe that in triangles AHC and CHB, $\angle AHC = \angle BHC = 90°$. So, in these two triangles we have equal angles and we have corresponding pairs of sides in the same ratio. Therefore, the triangles are similar. It implies that the other pairs of the respective angles must be equal, that is, $\angle HBC = \angle ACH$. Hence, it follows that

$$\angle ACB = \angle ACH + \angle HCB = \angle ACH + (90° - \angle HBC) = \angle ACH + (90° - \angle ACH) = 90°.$$

For Chapter 2, Problem 2.5

To prove that the median of a triangle separates it into two equal-area triangles.

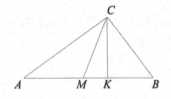

We consider triangle ABC with median CM and altitude CK. We have $AM = MB$ and $CK \perp AB$. Therefore, calculating the area of each triangle AMC and BMC as half the product of the base by the altitude dropped to the base we get

$$S_{AMC} = \frac{1}{2} AM \cdot CK = \frac{1}{2} BM \cdot CK = S_{BMC},$$

which is the sought-after result.

For Chapter 2, page 31: about **Poncelet–Steiner theorem.**
The ambitious readers can find more about this interesting topic in B. Pritsker, "*Geometrical Kaleidoscope*", Dover Publications, 2017.

For Chapter 3, Page 34

Property 3.5. The inverse image of a circle not passing through the center of inversion is a circle not passing through the center of inversion.

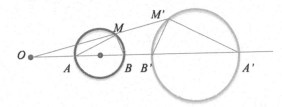

Proof.
Consider the inversion with center O and draw a straight line through O and the center of the given circle (the circle is not passing through O). Denote A and B the end points of the diameter of the given circle lying on this straight line. Pick an arbitrary point M on the circle other than A and B. Let A', B', and M' be the inverse images of A, B, and M, as shown in a diagram above. Denoting r the radius of the inversion, by definition of inversion, we have $OA \cdot OA' = OB \cdot OB' = OM \cdot OM' = r^2$. It follows that $\dfrac{OA}{OM} = \dfrac{OM'}{OA'}$ and $\dfrac{OB}{OM} = \dfrac{OM'}{OB'}$. We can establish two triangles similar if two pairs of corresponding sides are proportional and the included angles are congruent. This gives us the following pairs of similar triangles:

$$\Delta OAM \sim \Delta OM'A' \text{ and } \Delta OBM \sim \Delta OM'B'.$$

Therefore, the respective angles in these triangles are congruent, and we can state that

$$\angle MAO = \angle A'M'O \text{ and } \angle MBO = \angle B'M'O \qquad (A.1)$$

Now, we can observe, that since angle MAO is exterior angle of the triangle AMB, $\angle MAO = \angle AMB + \angle MBA$, and noticing that $\angle AMB = 90°$ (it is subtended by the diameter AB), we can rewrite the last equality as $\angle MAO = 90° + \angle MBA = 90° + \angle MBO$, from which

$$\angle MBO = \angle MAO - 90° \qquad (A.2)$$

Noticing that $\angle A'M'O = \angle B'M'O + \angle B'M'A'$, we can substitute the angle values from (A.1) and (A.2) to get that $\angle MAO = \angle MBO + \angle B'M'A' = (\angle MAO - 90°) + \angle B'M'A'$, from which $\angle B'M'A' = 90°$. It means that the inverse image of the right triangle AMB is right triangle $A'M'B'$, and points A', M', and B' lie on a circumcircle of the triangle $A'M'B'$ (inscribed angle $B'M'A'$ is subtended by the hypotenuse – diameter $B'A'$). Point M was selected arbitrarily on the given circle; therefore, the inverse image of that circle will be the circle passing through points A', B', and M' (where the hypotenuse $A'B'$ is its diameter). This concludes our proof that the inverse image of a circle not passing through the center of inversion is a circle not passing through the center of inversion.

For Chapter 3, Page 35

The *tangent-secant theorem* states that given a secant l intersecting the circle at points M and N and tangent t touching the circle at point T and given that l and t intersect at point P, the following equality holds: $PT^2 = PM \cdot PN$.

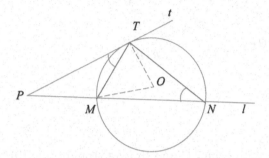

We will easily obtain the desired equality checking similar triangles PTN and PMT. These triangles are similar because they have a common angle at vertex P and $\angle MTP = \angle MNT$, which we will prove below.

Denote O the center of the given circle, and connect it with M and T. Triangle MOT is isosceles because $MO = TO$ as radii of the circle. Also, $OT \perp PT$ because it is given that PT is the tangent to the circle. It follows that

$$\angle MTP = 90° - \angle MTO = 90° - \frac{180° - \angle MOT}{2} = 90° - 90° + \frac{1}{2}\angle MOT = \frac{1}{2}\angle MOT. \quad \text{(A.3)}$$

On the other hand, applying the property of inscribed angles, we see that inscribed angle MNT which is subtended by the chord MT, equals half the corresponding central angle MOT,

$$\angle MNT = \frac{1}{2}\angle MOT. \quad \text{(A.4)}$$

Comparing (A.3) and (A.4), we see that indeed, $\angle MTP = \angle MNT$, which implies that $\Delta PTN \sim \Delta PMT$ (as we mentioned above, these triangles have the common angle at vertex P, and as proved, $\angle MTP = \angle MNT$, so the third respective angles must be congruent as well).

Hence, the respective sides in these triangles must be in proportion, namely, $\dfrac{PT}{PN} = \dfrac{PM}{PT}$, from which we derive the desired result, $PT^2 = PM \cdot PN$.

For Chapter 3, Page 37

The Pythagorean Theorem's proof using Ptolemy's Theorem.

In a right triangle ACB ($\angle C = 90°$) let's denote $BC = a$, $AC = b$, and $AB = c$. The goal is to prove that $a^2 + b^2 = c^2$.

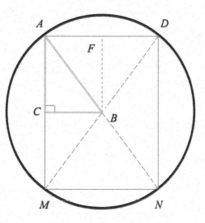

We draw a circle with center at B and radius $r = AB = c$. Extend AC till its intersection with this circle at M, extend AB till its intersection with circle at N, draw diameter MD (clearly, it passes through B), and connect D with A and N.

$\angle AMN = \angle MND = \angle ADN = \angle MAD = 90°$ as inscribed angles subtended by diameters. Therefore, as the result of our constructions, we got an inscribed rectangle $ADNM$. Draw $BF \perp AD$. Since $\triangle ADB$ is isosceles triangle ($AB = BD = r$), then BF being an altitude, is the median as well. Hence, $AF = FD = a$. Similarly, $AC = CM = b$.

Finally, $AN = MD = 2r$. Applying Ptolemy's Theorem, we have that $AM \cdot DN + AD \cdot MN = AN \cdot MD$, or in our nominations, $4b^2 + 4a^2 = 4c^2$. Cancelling out 4 on both sides gives the sought-after result, $a^2 + b^2 = c^2$.

For Chapter 4, Page 61, Proof 4.4

The Pitot theorem states that in a tangential quadrilateral the two sums of lengths of opposite sides are the same.

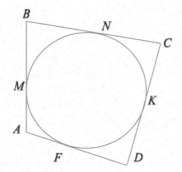

For the proof we will recall that two tangent line segments from a point outside a circle have equal lengths:

$AM = AF$, $BM = BN$, $CK = CN$, and $DK = DF$. Adding we obtain:

$AM + BM + CK + DK = AF + BN + CN + DF$, or $(AM + BM) + (CK + DK) = (AF + DF) + (BN + CN)$, which yields the desired result, $AB + CD = AD + BC$.

For Chapter 4, Page 63, Lemma

In our proof we used the assertion that is known as *Proposition 19* from *Euclid's Elements*:

A greater angle of a triangle is opposite a greater side.

Proposition 18 from the same work states:

A greater side of a triangle is opposite a greater angle.

Let's first prove Proposition 18.

Proof.

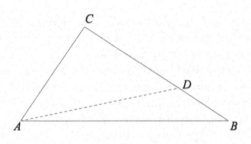

Consider triangle ABC in which BC is greater than AC. We need to show that angle CAB is greater than angle ABC.

Since $BC > AC$ then there exists point D on BC such $CD = AC$.

In isosceles triangle ACD, $\angle D = \angle A$. Considering triangle ADB, we see that the exterior angle ADC is greater than the opposite interior angle ABD (because $\angle ADC = \angle DAB + \angle ABD$).

Therefore, $\angle CAD = \angle ADC > \angle ABD$. Notice that $\angle CAB = \angle CAD + \angle DAB$, so, clearly, $\angle CAB > \angle CAD = \angle ADC$. It implies that, as requested, $\angle CAB > \angle ABD = \angle ABC$, and our proof is completed.

Now we will prove Proposition 19.

We can use the same diagram as above. Here we are given that angle CAB is greater than angle ABC, and we need to prove that BC is greater than AC.

Assume the converse that BC is either equal to AC or less.

If $BC = AC$ then triangle ABC is isosceles and $\angle CAB = \angle ABC$, which contradicts to the given condition that $\angle CAB > \angle ABC$. So, it is impossible that $BC = AC$. If we assume now that $BC < AC$ than by earlier proved Proposition 18, $\angle CAB < \angle ABC$, which also contradicts the given condition that $\angle CAB > \angle ABC$. Therefore, in either case, we get to the contradiction with the given condition. This proves our assertion that indeed, a greater angle of a triangle is opposite a greater side.

For Chapter 4, Page 64

Theorem.
Each angle bisector of a triangle divides the opposite side into segments proportional in length to the adjacent sides.

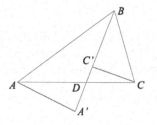

Proof.
In triangle ABC, BD is the bisector of angle B. We have to prove that $\dfrac{AD}{AB} = \dfrac{DC}{BC}$. Let us draw $AA' \perp BD$ (its extension) and $CC' \perp BD$. Triangles $AA'D$ and $CC'D$ are similar, because $\angle AA'D = \angle CC'D = 90°$ and $\angle ADA' = \angle CDC'$ as vertical angles.

Therefore,

$$\frac{AD}{AA'} = \frac{DC}{CC'}. \tag{A.5}$$

Triangles $AA'B$ and $CC'B$ are also similar, because $\angle ABA' = \angle CBC'$ (it is given that BD divides angle ABC in half) and $\angle AA'B = \angle CC'B = 90°$, respectively the remaining pair of angles in each triangle are equal as well.

Hence,

$$\frac{AB}{AA'} = \frac{BC}{CC'}. \tag{A.6}$$

Dividing (A.5) by (A.6) yields the desired result, $\dfrac{AD}{AB} = \dfrac{DC}{BC}$.

For Chapter 4, Page 68

Several proofs that MN is the quadratic mean of the bases of a trapezoid (the most difficult of all cases, in our view) were discussed in B. Pritsker "Geometrical Kaleidoscope", Dover Publications Inc., 2017 and B. Pritsker "Mathematical Labyrinths. Pathfinding", World Scientific, 2021.

There are several other useful tools hidden in the literature which may also be of value to ambitious readers. For example, I can refer readers to very interesting article "Remarkable limits" by M. Crane and A. Nudelman published in *Quantum*, July/August 1997 issue. It covers calculations of limits using classic means. Amazingly, some calculations, for instance, Schwab-Schenberg mean, were derived from elementary geometrical considerations.

For Chapter 5, Page 70, Problem 5.1

Problem 5.1 solved in Chapter 5 presents the proof of the necessary conditions of *Carnot's Theorem* (named after a French mathematician Lazare Nicolas Marguerite, Count Carnot (1752–1823)).

Direct Statement

For a triangle ABC consider three perpendiculars to the sides that intersect in a common point P, $l_1 \perp BC$, $l_2 \perp AB$, $l_3 \perp AC$. If M, N, and K are the pedal points of those three perpendiculars on the sides AB, BC, and CA, then the following equality holds:

$$AM^2 + BN^2 + CK^2 = MB^2 + NC^2 + AK^2.$$

Converse Statement

If the above equality holds for the pedal points of three perpendiculars on the three triangle sides, then they intersect in a common point.

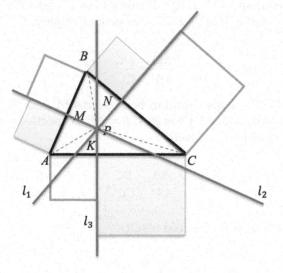

Let's prove the converse statement.

Consider the point P such that $PN \perp BC$ and $PM \perp AB$, and draw $PF \perp AC$ ($F \in AC$). Our goal is to show that under the given conditions, F coincides with K. Assume that $F \neq K$. Then by the direct theorem, the following equality holds true $AM^2 + BN^2 + CF^2 = MB^2 + NC^2 + AF^2$.

Subtracting this equality from the given $AM^2 + BN^2 + CK^2 = MB^2 + NC^2 + AK^2$, we get that $CK^2 - CF^2 = AK^2 - AF^2$. This can be written as $CK^2 - AK^2 = CF^2 - AF^2$ and further modified to $(CK - AK)(CK + AK) = (CF - AF)(CF + AF)$.

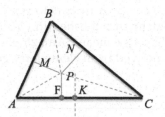

Noticing that $CK + AK = CF + AF = AC$ and dividing both sides by AC, finally yields $CK - AK = CF - AF$ or equivalently, $CK + AF = CF + AK$ (*).

In assumption that $F \neq K$, it is either F lies between A and K, as we have it in the diagram above, or K lies between A and F. If F lies between A and K, then $AF < AK$ and $CK < CF$.

This implies that $CK + AF < CF + AK$ which contradicts to (*). We will get similar contradiction in case when K lies between A and F. Therefore, K and F must coincide proving our assertion that if the given equality holds for the pedal points of three perpendiculars on the three triangles' sides, then they intersect at a common point.

For Chapter 5, Page 72

Theorem of Three Perpendiculars.
Theorem: If AB is perpendicular to a plane σ and if from B, the foot of the perpendicular, a straight line BC is drawn perpendicular to any straight line ST in the plane σ, then AC is also perpendicular to ST.

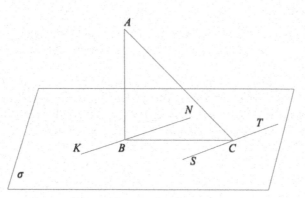

First, draw at B in the plane σ straight line KN parallel to ST. Since $KN \parallel ST$ and $BC \perp ST$ then $BC \perp KN$. We know that AB is perpendicular to the plane σ, hence AB is perpendicular to any straight line lying in σ, and it follows that $AB \perp KN$. But now we see that KN is perpendicular to the two intersecting straight lines BC and AB in the plane (ABC), which means that KN is perpendicular to the plane (ABC). Now, ST and KN are parallel, and KN is perpendicular to the plane (ABC); hence, ST is perpendicular to the plane (ABC) as well. Therefore, ST is perpendicular to AC (as to any straight line in the plane (ABC)), or in other words, AC is perpendicular to ST, as we wanted to prove.

For Chapter 5, Page 79

Viete's formulas for a quadratic equation $ax^2 + bx + c = 0,\ (a \neq 0)$

$$x_1 + x_2 = -\frac{b}{a},\tag{A.7}$$

$$x_1\ x_2 = \frac{c}{a}.\tag{A.8}$$

Viète's Theorem

Direct Statement

If x_1 and x_2 are the roots of a quadratic equation $ax^2 + bx + c = 0$, then formulas (A.7) and (A.8) hold true.

Converse Statement

The numbers x_1 and x_2 are the roots of a quadratic equation $x^2 - (x_1 + x_2) x + x_1 x_2 = 0$.

For Chapter 5, Page 82

We can suggest Mario Livio, *"The Golden Ratio: The Story of Phi, the World's Most Astonishing Number"*, Broadway Books, 2002 and Nikolai Vorob'ev, *"Fibonacci Numbers"*, Dover Publications Inc., 2011.

Another two sources for information on interaction of the Golden Ratio and Fibonacci numbers specifically related to solutions of the Diophantine equations and geometrical paradoxes can be found in B. Pritsker, *"The Equations World"*, Dover Publications, 2019 and B. Pritsker, *"Mathematical Labyrinths. Pathfinding"*, World Scientific, 2021.

For Chapter 5, Pages 85 and 86

Proofs of the formulas for the area of a convex quadrilateral and their applications in problem solving can be found in B. Pritsker, *"Geometrical Kaleidoscope"*, Dover Publications, 2017.

For Chapter 5, Pages 86, Euler's Formula

Euler's formula gives a relation between the distance d between the circumcenter and incenter of a triangle and the lengths of the radii of the circumcircle and incircle, R and r respectively.

The formula $d^2 = R^2 - 2Rr$ is named after Leonhard Euler, a Swiss mathematician, physicist, logician, astronomer, and engineer, who published it in 1765. The same result was published earlier by William Chapple (1718–1781), an English mathematician, in 1746. In some texts this formula is written as $\dfrac{1}{R-d} + \dfrac{1}{R+d} = \dfrac{1}{r}$ or $(R-r)^2 = d^2 + r^2$.

As we explored the notion of inversion and devoted the entire chapter to its properties, we will use this opportunity to mention a very interesting and relatively simple proof of Euler's formula that is based on the inversion properties. This stunning proof is presented at Alexander Bogomolny's "Cut-the-knot" web site at www.cut-the-knot.org, and it is credited to Nathan Bowler.

Before we proceed to the proof of the formula, let's introduce the following Lemma (see the figure below):

> *If three circles having the same radius pass through a point, the circle through their other three*
> *points of intersection also has the same radius.*

We have three circles with the centers O_1, O_2, and O_3 that have the same radii passing through at O. Denote the other points of their intersections A, B, and C. Since all circles have equal radii, then OO_2BO_3 and OO_1AO_3 are congruent rhombuses with the common side OO_3. Therefore, in a quadrilateral AO_1O_2B the sides AO_1 and BO_2 are parallel and have the same length. It implies that AO_1O_2B is a parallelogram, and respectively, $O_1O_2 = AB$. In a similar fashion we can prove that $O_1O_3 = CB$ and $O_2O_3 = CA$.

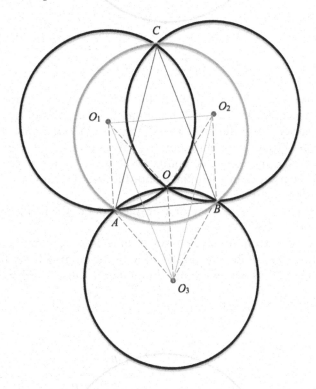

Therefore, the triangles ABC and $O_1O_2O_3$ are congruent by three sides, and have equal circumcircles. But the circumcircle of the former has its center at O (this is the common point of intersection of three circles that is equidistant from all three centers), and is equal to each of the given circles. Hence the orange circle through A, B, and C is equal to each of the given circles, which completes the proof of the lemma.

We now pass to the proof of the formula $d^2 = R^2 - 2Rr$.

The idea underlying the proof is based on the following theorem:

> *The inverse images of the sides of a triangle in its incircle are three circles of equal radii that con-*
> *cur at the incenter. The circle through their second points of intersection is the inverse image of*
> *the circumcircle. It has the same radius.*

For the given triangle ABC, we consider the inversion in its incircle (see Figure A.1).

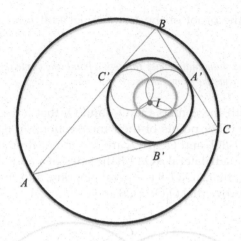

FIGURE A.1

Since the sides of our triangle are tangent to the incircle, they will invert into the circles passing through the incenter I, which will be tangent to the sides of $\triangle ABC$ at some points A', B', and C'. All three circles will have equal diameters $IA' = IB' = IC'$. The second points of intersection of these circles correspond by the inversion to the vertices of $\triangle ABC$. It follows that the circle through these three points is the inverse image of the circumcircle of $\triangle ABC$, and as proved earlier in our lemma, it should have the same diameter as the three mentioned circles (orange circle).

With this introduction, the stage is set for a simple proof of the formula. Figure A.2 demonstrates the idea of the proof.

Denote O the center of the circumcircle of $\triangle ABC$ and draw a straight line through O and I. Denote P and Q the points of intersection of OI with the circumcircle, and P' and Q' their respective inversion images. Then in our nominations, $OI = d$, $IA' = IB' = IC' = Q'P' = r$, and $OA = OB = OC = OP = OQ = R$.

Also, we see that $IP = OP - OI = R - d$ and $IQ = OQ + OI = R + d$.

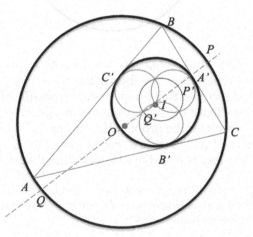

FIGURE A.2

By definition of the inversion, $IP \cdot IP' = IQ \cdot IQ' = r^2$, from which $IP' = \dfrac{r^2}{IP}$ and

$IQ' = \dfrac{r^2}{IQ}$.

Noticing that $IP' + IQ' = r$ and substituting the respective expressions into this equality gives $\dfrac{r^2}{IP} + \dfrac{r^2}{IQ} = r$ or equivalently,

$\dfrac{r^2}{R-d} + \dfrac{r^2}{R+d} = r$, and after a few simplifications we arrive at

$$(R-d)(R+d) = r(R+d+R-d),$$
$$R^2 - d^2 = 2Rr,$$

$d^2 = R^2 - 2Rr$, which is the desired result.

Speaking about the geometrical inequalities, it worth mentioning that if we rewrite Euler's formula as $R^2 = d^2 + 2Rr$, it immediately follows that $R^2 \geq 2Rr$, and after dividing both sides by R, we obtain $R \geq 2r$, which is the famous *Euler inequality* (the equality holds only for an equilateral triangle) relating the radii of the circumscribed and the inscribed circles.

For Chapter 6, Trigonometric Relations

Sum and Difference Formulas

$$\sin(x+y) = \sin x \cos y + \sin y \cos x;$$

$$\sin(x-y) = \sin x \cos y - \sin y \cos x;$$

$$\cos(x+y) = \cos x \cos y - \sin x \sin y;$$

$$\cos(x-y) = \cos x \cos y + \sin x \sin y;$$

$$\tan(x-y) = \frac{\tan x - \tan y}{1 + \tan x \tan y};$$

$$\tan(x+y) = \frac{\tan x + \tan y}{1 - \tan x \tan y};$$

$$\cot(x+y) = \frac{\cot x \cot y - 1}{\cot y + \cot x};$$

$$\cot(x-y) = \frac{\cot x \cot y + 1}{\cot y - \cot x}.$$

Double-Angle Formulas

$$\sin 2x = 2\sin x \cos x;$$

$$\cos 2x = \cos^2 x - \sin^2 x;$$

$$\tan 2x = \frac{2\tan x}{1 - \tan^2 x};$$

$$\cot 2x = \frac{\cot^2 x - 1}{2\cot x}.$$

Sum-to-Product Formulas

$$\sin x + \sin y = 2\sin\frac{x+y}{2}\cos\frac{x-y}{2};$$

$$\sin x - \sin y = 2\sin\frac{x-y}{2}\cos\frac{x+y}{2};$$

$$\cos x + \cos y = 2\cos\frac{x+y}{2}\cos\frac{x-y}{2};$$

$$\cos x - \cos y = 2\sin\frac{x+y}{2}\sin\frac{y-x}{2} = -2\sin\frac{x+y}{2}\sin\frac{x-y}{2};$$

$$\tan x \pm \tan y = \frac{\sin(x\pm y)}{\cos x \cos y};$$

$$\cot x \pm \cot y = \frac{\sin(y\pm x)}{\sin x \sin y}.$$

Product-to-Sum Formulas

$$\sin x \sin y = \frac{1}{2}\left(\cos(x-y) - \cos(x+y)\right);$$

$$\sin x \cos y = \frac{1}{2}\left(\sin(x+y) + \sin(x-y)\right);$$

$$\cos x \cos y = \frac{1}{2}\left(\cos(x-y) + \cos(x+y)\right);$$

Power-Reducing Formulas

$$\sin^2 x = \frac{1}{2} - \frac{1}{2}\cos 2x;$$

$$\cos^2 x = \frac{1}{2} + \frac{1}{2}\cos 2x.$$

Half-Angle Formulas

$$\left|\sin\frac{x}{2}\right| = \sqrt{\frac{1-\cos x}{2}};\ \left|\cos\frac{x}{2}\right| = \sqrt{\frac{1+\cos x}{2}};\ \left|\tan\frac{x}{2}\right| = \sqrt{\frac{1-\cos x}{1+\cos x}};\ \left|\cot\frac{x}{2}\right| = \sqrt{\frac{1+\cos x}{1-\cos x}}.$$

$$\tan\frac{x}{2} = \frac{\sin x}{1+\cos x} = \frac{1-\cos x}{\sin x};\ \cot\frac{x}{2} = \frac{1+\cos x}{\sin x} = \frac{\sin x}{1-\cos x}.$$

Universal Trigonometric Substitution

$$\sin x = \frac{2\tan\frac{x}{2}}{1+\tan^2\frac{x}{2}};\ \cos x = \frac{1-\tan^2\frac{x}{2}}{1+\tan^2\frac{x}{2}};\ \tan x = \frac{2\tan\frac{x}{2}}{1-\tan^2\frac{x}{2}}.$$

Cofunction Identities

$$\sin\left(\frac{\pi}{2}-x\right) = \cos x,\ \cos\left(\frac{\pi}{2}-x\right) = \sin x,\ \tan\left(\frac{\pi}{2}-x\right) = \cot x,\ \cot\left(\frac{\pi}{2}-x\right) = \tan x.$$

For Chapter 7, Problem 7.5 Page 120

Let's prove that in a regular triangle, the circumradius is expressed in terms of its side as $R = \dfrac{a}{\sqrt{3}}$.

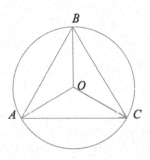

Denote $AB = BC = AC = a$ and $OA = OB = OC = R$.

Applying the Law of cosines to triangle BOC gives

$$BC^2 = BO^2 + CO^2 - 2BO\cdot CO\cdot\cos\angle BOC,\quad \text{or in our nominations, noticing that}$$

$\angle BOC = 120°$, we get that $a^2 = R^2 + R^2 - 2R^2\cdot\left(-\dfrac{1}{2}\right) = 3R^2$. Thus, $R = \dfrac{a}{\sqrt{3}}$, as we wished to prove.

For Chapter 7, Page 128

Intersecting Chords Theorem states that if two chords intersect in a circle, the product of the lengths of the segments of one chord equal the product of the segments of the other, i.e. $AS \cdot SC = BS \cdot SD$, and conversely, if for two line segments AC and BD intersecting in S the equation above holds true, then all four points A, B, C, and D lie on a common circle ($ABCD$ is a cyclic quadrilateral).

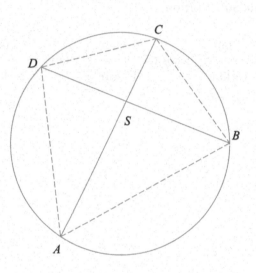

Proof – Direct Statement.
Chords AC and BD intersect at S. We need to prove that $AS \cdot SC = BS \cdot SD$.
Checking inscribed angles subtended by the same chords, we see that
$\angle ADB = \angle BCA$ (subtended by AB), or equivalently, $\angle ADS = \angle BCS$.
$\angle DAC = \angle CBD$ (subtended by CD), or equivalently, $\angle DAS = \angle CBS$.
Also, $\angle ASD = \angle CSB$ as vertical angles. Therefore, triangles ASD and BSC are similar. It follows that $\dfrac{AS}{SD} = \dfrac{BS}{SC}$, which can be written as $AS \cdot SC = BS \cdot SD$, the sought-after result.

Proof – Converse Statement.
Here, we are given two line segments AC and BD intersect at S and $AS \cdot SC = BS \cdot SD$. We need to prove that $ABCD$ is a cyclic quadrilateral, i.e. all four points, A, B, C, and D lie on a common circle.
   Let's draw a circle through points A, B, and C (clearly, these points are not collinear because we know that AC and BD intersect at S) and prove that D must lie on the same circle as well. Assume this circle intersect extension of BS at D'. Then we can apply the above proof of the direct statement to get $AS \cdot SC = BS \cdot SD'$. Comparing this equality to the given equality that $AS \cdot SC = BS \cdot SD$, we conclude that D and D' coincide. Therefore, indeed, all four points lie on the same circle.

For Chapter 8, Problem 8.12, Page 144

Alternative way of solving this problem is to calculate the complementary probability of rolling not more than 4 in the first and second toss: $p = 1 - \left(\dfrac{2}{3}\right)^2 = 1 - \dfrac{4}{9} = \dfrac{5}{9}$.

For Chapter 9, Problem 9.6, Page 151

The volume of a parallelepiped is the product of the base area and the height:
$V = S_{base} \cdot h$. The volume of the rectangular parallelepiped respectively is the product of the area of a rectangle (its base) and the height.

For Chapter 10, Problem 10.10, Page 174

The Intercept Theorem or Thale's Theorem states that if two or more parallel lines are intersected by two self-intersecting lines, then the ratios of the line segments of the first intersecting line is equal to the ratio of the similar line segments of the second intersecting line.

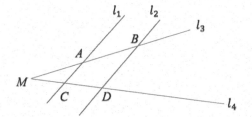

It is given that $l_1 \parallel l_2$ and l_3 intersects l_4 at point M. Denote the points of intersection of l_1 and l_2 with l_3 and l_4 respectively A, B, C, and D.

We need to prove that $\dfrac{MA}{AB} = \dfrac{MC}{CD}$.

Checking angles, we see that triangles MAC and MBD are similar.

It implies that $\dfrac{MA}{MB} = \dfrac{MC}{MD}$ (*).

Using this equality we can easily prove the theorem's assertion. To simplify calculations, we introduce several variables: $MA = x$, $AB = y$, $MC = z$, and $CD = t$.

We have $MB = x + y$ and $MD = z + t$. Substituting these expressions into (*) gives $\dfrac{x}{x+y} = \dfrac{z}{z+t}$, from which $x(z+t) = z(x+y)$. Simplifying the last equality yields $xz + xt = zx + zy$, and finally, $xt = zy$. The last equality can be rewritten as $\dfrac{x}{y} = \dfrac{z}{t}$, or making reverse substitutions, we get the desired result $\dfrac{MA}{AB} = \dfrac{MC}{CD}$.

For Chapter 10, Problem 10.15, Page 181

Carefully drawing segments, one should get that *MO* and *NO* will not intersect inside of *ABDC*. Moreover, the point of their intersection *O* will be such that there will not be any equal angles formed to get their sums to be equal to ∠*CAB* and ∠*DBA*, as we referred to in our proof, misleading the readers!

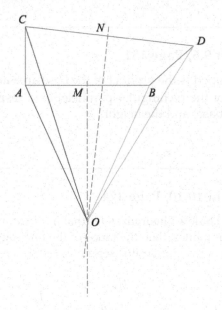

References

A. Bogomolny, *Cut-the-knot*, Educational website, http://www.cut-the-knot.org/content.shtml#

A. Yegorov, *"Inequalities become equalities"*, Quantum, March/April, 2000

A. Yegorov, *"The discriminant at work"*, Quantum, January/February, 1996

B. Hofmann-Wellenhof, K. Legat, M. Wieser, H. Lichtenegger, *Navigation: Principles of Positioning and Guidance*. Springer, 2003

B. Pritsker, *"Geometrical Kaleidoscope"*, Dover Publications, 2017.

B. Pritsker, *"Mathematical Labyrinths. Pathfinding"*, World Scientific, 2021.

B. Pritsker, *"The Equations World"*, Dover Publications, 2019.

Boris A. Kordemsky, *"The Moscow Puzzles"*, Dover Publications, 2016

Bruen, A., Fischer, J.C. and Wilker, J.B. *Apollonius by inversion*. Mathematics Magazine (56) 1983, 97–103.

Coxeter, H. S. M. and S. L. Greitzer, *"Geometry Revisited"*, Mathematical Association of America, 1967

E. Vinberg, *"A revolution absorbed"*, Quantum, January/February, 1997

Faber, Richard L., *Foundations of Euclidean and Non-Euclidean Geometry*, New York: Marcel Dekker Inc., 1983

G. Polya, *Mathematical Discovery*, John Wiley and Sons, 1981

Hardy, G. H., Littlewood, J. E., and Polya, G., *Inequalities*, Cambridge Univ. Press, 1952

Julia Angwin *"What is elegance?"*, Quantum, January/February 1995

M. Apresyan, *"Infinite algebraic tilings"*, Quantum, May/June 1994

M. Saul, T. Andreescu, *"Symmetry in algebra"*, Quantum, March/April, 1998

"Mathematical Puzzles of Sam Loyd" selected and edited by Martin Gardner, Dover Publications, 2017

N. Vasilyev, *"The Symmetry of chance"*, Quantum, May/June, 1993.

R. Courant and H. Robbins, *What is Mathematics?*, Oxford University Press, 1996

R. Hartshorne, *Geometry: Euclid and Beyond*, Springer, 2000

R. Honsberger, *Mathematical Gems II*, MAA, 1976

S. Ovchinnikov and I. Sharygin, *"Numerical Data in geometry problems"*, Quantum, May/June 1999

V. Boltyansky, *"Turning the Incredible into the obvious"*, Quantum, September/October, 1992.

V. N. Beryozin, *"The Good old Pythagorean Theorem"*, Quantum, January/February, 1994

В. Болтянский, Ю. Сидоров, М. Шабунин, *"Лекции и задачи по элементарной математике"*, Москва, Наука, 1972 (in Russian)

В. Михайловський, М. Ядренко, Г. Призва, В. Вишенський, *"Збірник задач республіканських математичних олімпіад"*, Київ, Вища школа, 1979 (in Ukrainian)

В. Чистяков, *"Старинные задачи по элементарной математике"*, Минск Вышэйшая школа, 1978 (in Russian)

С. Лавренов, *"Множество значений функции"*, Квант, #4, 2007 (in Russian)

Е. Игнатьев, *"В Царстве Смекалки"*, Москва, Наука, 1984 (in Russian)

Я. Перельман, *"Занимательная Алгебра. Занимательная Геометрия"*, Москва АСТ, 1999 (in Russian)

Я. Суконник, *"Математические задачи повышенной трудности"*, Киев, Радянська Школа, 1985 (in Russian)

Сергев И.Н., *"Зарубежные Математические Олимпиады"*, Москва, Наука, 1987 (in Russian)

Э. Готман, З. Скопец, *"Задача одна - решения разные"*, Киев, Радянська Школа, 1988 (in Russian)

А. Анджанс, Д, Бонка, *"Метод интерпретаций"*, Квант, #1, 2009 (in Russian)

Э. Готман, З. Скопец, *"Решение Геометрических Задач Аналитическим Методом"*, Москва, Просвещение, 1979 (in Russian)

М. Крайзман, *"Заменим фигуру"*, Квант, #5, 1979 (in Russian)

З. А. Скопец, *"Сравнение различных средних двух положительных чисел"*. Квант, #2, 1971 (in Russian)

Гольдман А., Звавич Л., *"Числовые средние и геометрия"*. Квант, #9, 1990 (in Russian)

Затакавай В., *"Теорема Птоломея и некоторые тригонометрические соотношения"*. Квант, #4, 1991 (in Russian)

М. И. Сканави, *"Сборник конкурсных задач по математике для поступающих во втузы"*, Москва, Высшая Школа, 1980 (in Russian)

Ш. Горделадзе, М. Кухарчук, Ф. Яремчук, *"Збірник Конкурсних Задач з Математики"*, Київ, Вища Школа, 1988 (in Ukrainian)

Index